SpringerWienNewYork

Th. Müller and P. Riederer (eds.)

Focus on Extrapyramidal Dysfunction

Journal of Neural Transmission
Supplement 68

SpringerWienNewYork

Prof. Dr. Th. Müller
Neurologische Klinik, St. Josef Hospital
Gudrunstraße 56
44791 Bochum
Germany

Prof. Dr. P. Riederer
Klinik und Poliklinik für Psychiatrie und Psychotherapie
Füchsleinstraße 15
97080 Würzburg
Germany

Printed in Austria

SpringerWienNewYork is a part of Springer Science+Business Media
springeronline.com

Typesetting: Thomson Press India Ltd., New Dehli
Printing: Holzhausen Druck & Medien GmbH, 1140 Wien

Printed on acid-free and chlorine-free bleached paper

SPIN: 10990292

Library of Congress Control Number: 2004109707

With numerous Figures

ISSN 0303-6995
ISBN 3-211-21114-4 (soft cover) SpringerWienNewYork
ISBN 3-211-21112-8 (hard cover) SpringerWienNewYork

Preface

We present this volume in honour of Professor Dr. med. Horst Przuntek with our best wishes on the occasion of his 65th birthday. Horst Przuntek has occupied the position of head of the Department of Neurology at St. Josef Hospital of Ruhr University at Bochum in Germany since 1986. He built up this Department into one of the largest and most highly-regarded in all the German neurological hospitals. We wish him continued success in his work over the next years.

Horst Przuntek created the ideal conditions for the establishment of basic and predominantly clinical research in various fields of neurology, and thereby established for himself a leading and influential position among neurologists. His wealth of ideas combined with enthusiasm and critical judgement, good-natured and inventive motivation, and the occasional visionary and provocative suggestions have created a stimulating and friendly working environment and a co-operative spirit among his colleagues. He is particularly welcoming of, and always ready to accept, novel developments at the cutting edge of therapeutic advances in the field of neurological disorders. In his clinical practice he insists on early and rapid diagnosis based on a careful taking of the patient's history, followed by investigations carried out with the efficient use of the necessary techniques, all of which he has developed and constantly improved by his pragmatic and predictive interdisciplinary approach and powers of persuasion, as became evident soon after his arrival at St. Josef Hospital.

Horst Przuntek is devoted to providing the most favourable conditions for the work of his department, requiring the best and most carefully designed treatment for the patients in his care based on the latest findings, from his personal vision of the role of the physician. His determination is to understand the patient's disease and its appearance in the context of the entire body and its interaction with the patient's whole personality. This more comprehensive view of life is the key to his passionate aspiration to help his patients and not only to the diagnosis of their disease. This more global appreciation of disease processes, combined with his extensive personal clinical experience, is the underlying basic condition for the successful design, execution and interpretation of his widely accepted pioneering long-term trials of treatments for Parkinson's disease. Horst Przuntek has used not only the current medicines in the treatment of Parkinson's disease, but has also contributed to clinical scientific studies on experimental substances, as illustrated in the three appended references of his widely cited studies with selegiline, bromocriptine, and budipine.

Horst Przuntek's scientific work has been substantial, comprising over 470 publications on fundamental and clinical-scientific themes, beginning in 1967 with experimental studies on the intra-vesicular storage of noradrenaline. This was the beginning of an ever widening span of neurological studies with a strong focus on extrapyramidal disorders, particularly Parkinson's disease. In accordance with his strong motivation towards the development of creative and forward-looking concepts, he has recently joined the two of us in putting the neuropathological concept of Braak on the staging of investigations of Parkinson's disease on a corresponding clinical basis according to our current state of knowledge.

Horst Przuntek's personality and his exemplary demeanor as a physician contribute to making him an experienced, successful, honoured and popular – but also at times unconventional – teacher because of the way the is able to convey his own excitement in presenting neurology, neuropsychiatry and geriatrics at his high-level academic lectures.

He has been the godfather and a founder member of the International Graduate School for Neuroscience at the Ruhr University at Bochum from an early realisation of how the combination of basic and clinical aspects of research in the interdisciplinary fields of neurobiology, neurology and psychiatry were and would continue to be one of the most striking challenges in the field of medicine and clinical patient care. In line with these concepts, Horst Przuntek has organised several symposia at which neurologists and psychiatrists have focused on these scientific topics from their own individual viewpoints.

Close colleagues and previous and present staff members join together in presenting their recent findings in this volume, at least partially reflecting the enormous variety of Horst Przuntek's scientific approach, and these presentations represent the fruits of his ongoing never-ending wealth of ideas and efforts in his quest for a better understanding of disease.

He writes in one of his poems *"I could never be alone ..."*, and perhaps this is the clue to his continued need for interaction, discussion and engagement in all the fascinating aspects of life.

Bochum/Würzburg, June 2004

Th. Müller
P. Riederer

References

Przuntek H, Bittkau S, Bliesath H, Buttner U, Fuchs G, Glass J, Haller H, Klockgether T, Kraus P, Lachenmayer L, Muller D, Muller T, Rathay B, Sgonina J, Steinijans V, Teshmar E, Ulm G, Volc D (2002) Budipine provides additional benefit in patients with Parkinson disease receiving a stable optimum dopaminergic drug regimen. Arch Neurol 59: 803–806

Przuntek H, Conrad B, Dichgans J, Kraus PH, Krauseneck P, Pergande G, Rinne U, Schimrigk K, Schnitker J, Vogel HP (1999) SELEDO: a 5-year long-term trial on the effect of selegiline in early Parkinsonian patients treated with levodopa. Eur J Neurol 6: 141–150

Przuntek H, Welzel D, Gerlach M, Blumner E, Danielczyk W, Kaiser HJ, Kraus PH, Letzel H, Riederer P, Uberla K (1996) Early institution of bromocriptine in Parkinson's disease inhibits the emergence of levodopa-associated motor side effects. Long-term results of the PRADO study. J Neural Transm 103: 699–715

Contents

Listed in Current Contents/Life Sciences

J Neural Transm (2004) [Suppl] 68: 1–13

Dopaminergic neurotoxicity of homocysteine and its derivatives in primary mesencephalic cultures

I. Heider[1], **V. Lehmensiek**[1], **Th. Lenk**[1], **Th. Müller**[2], and **A. Storch**[1]

[1] Department of Neurology, University of Ulm, and
[2] Department of Neurology, Ruhr University of Bochum, Germany

Summary. Levodopa and dopamine are metabolized to 3-*O*-methyldopa and 3-methoxytyramine, respectively, by the enzyme catechol-*O*-methyltransferase (COMT) leading to the production of the demethylated cofactor *S*-adenosylhomocysteine (SAH) and subsequently homocysteine (HC). Indeed, treatment of Parkinson's disease (PD) patients with levodopa leads to increased HC blood levels. Therefore, HC is discussed to be involved in the pathogenesis of PD as well as in enhanced progression of PD in patients treated with levodopa. Here we investigated the toxicity of HC and its derivatives SAH, homocysteic acid (HCA) and cysteic acid (CA) on tyrosine hydroxylase (TH)-positive neurons in primary mesencephalic cultures from rat *in vitro*. Furthermore, we evaluated the toxicity of HC on cultures stressed with the dopaminergic neurotoxin 1-methyl-4-phenylpyridinium (MPP^+). Incubation with HC or HCA did not result in significant effects on TH-positive neuron survival with concentrations up to 1 mM, but led to morphological changes of TH-positive cells with significantly fewer and shorter neurites at concentrations of $\geq$100 μM after 48 h. In contrast, SAH and CA were toxic at concentrations of >100 μM after 48 h. Furthermore, MPP^+ showed strong toxicity towards TH-positive cells after 48 h (half-maximal toxic concentration: 20 μM), whereas co-incubation with HC for 24 or 48 h did not further alter TH-positive cell survival. Taken together, our results do not demonstrate relevant dopaminergic toxicity of HC *in vitro*, and therefore HC is most likely not involved in the pathogenesis of PD or in accelerating the progression of PD by levodopa.

Introduction

Parkinson's disease (PD) is characterized by dysfunction and degeneration of dopaminergic neurons in the *substantia nigra pars compacta* resulting in decreased dopamine levels in the striatum subsequently leading to progressive akinesia, tremor and rigidity (Lang and Lozano, 1998). The metabolic precursor of dopamine, blood-brain trespassing levodopa, is one of the most effective antiparkinsonian drug, but there are increasing evidences for an accelerating action of levodopa on disease progression of PD (Blunt et al., 1993; Fahn et al., 2002; Luquin et al., 2001). However, levodopa is usually administered with dopadecarboxylase (DDC) inhibitors to prevent peripheral degradation to dopamine (Tolosa et al., 1998). This increases the conversion of levodopa to 3-*O*-methyldopa by catechol-*O*-methyltransferase (COMT) in blood, peripheral tissue and nigrostriatal neurons (Fig. 1; Miller et al., 1997; Wurtman et al., 1972; Yassin et al., 1998). COMT catalyzes the transfer of a methyl group from the co-factor *S*-adenosyl-L-methionine (SAM) to the phenolic group of catecholic compounds, such as levodopa (Jeffrey and Roth, 1984; Männistö et al., 1992). Thus, *O*-methylation of levodopa is associated with conversion of SAM into *S*-adenosylhomocysteine (SAH) and subsequently homocysteine (HC; Fig. 1; Miller et al., 1997; Wurtman et al., 1972; Yassin et al., 1998). Indeed, recent studies showed that HC levels are increased in PD patients treated with levodopa compared to matched controls or previously untreated PD patients (Allain et al., 1995; Kuhn et al., 1998; Müller et al., 1999; Yasui et al., 2000). HC levels in treated PD patients are similar to those measured in mild hyperhomocysteinemia ($\approx$15–50 μM) (Lindgren et al., 1995; Perry et al., 1995). Synaptic levels of HC are unknown, but cerebral HC concentrations were reported to be up to 10 μM (Broch et al., 1984), whereas cerebrospinal fluid levels are $\approx$0.5 μM (Selhub et al., 1995). HC plasma levels are correlated to the total daily dose of levodopa (Allain et al., 1995; Kuhn et al., 1998; Yasui et al., 2000). It therefore remains unclear whether HC plays a critical role in the pathogenesis of PD and/or in accelerating the progression of PD in patients treated with levodopa.

Interestingly, inhibition of the conversion of levodopa into 3-*O*-methyldopa by blocking COMT activity *in vitro* leads to significant protection of dopaminergic neurons against levodopa toxicity (Storch et al., 2000a; Blessing et al., 2003). Extensive studies to elucidate the mechanisms suggest that levodopa toxicity towards dopaminergic neurons is mediated by an autooxidative process, which is attenuated by glial cells (Blessing et al., 2003; Mena et al., 1993, 1997). In addition, we demonstrated a second pathway of levodopa toxicity *in vitro* mediated by a COMT- and glia-dependent pathway (Blessing et al., 2003). In this

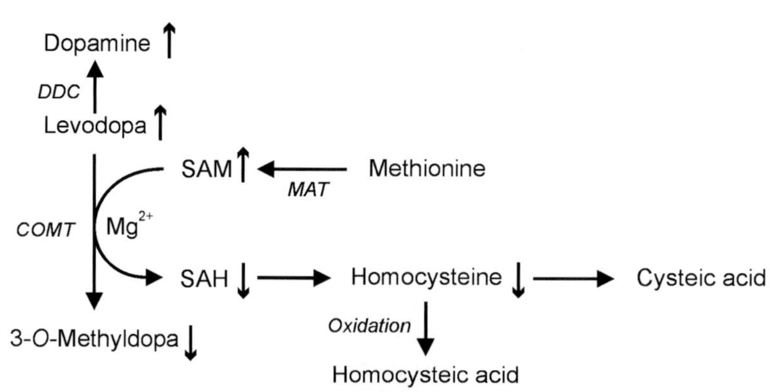

Fig. 1. Metabolic pathways of levodopa with respect to the methyl group metabolism. Please note that the conversion of homocysteine into cysteic acid includes several metabolic steps. Closed arrows represent metabolic pathways including enzymes (in *italic*), open arrows represent changes in the levels of the respective metabolite in the case of COMT inhibition (see text for details). *SAM* *S*-adenosyl-L-methionine; *SAH* *S*-adenosylhomocysteine; *COMT* catechol-*O*-methyltransferase; *DDC* dopadecarboxylase; *MAT* methionine-adenosyltransferase

pathway, one possible protective mechanism includes enhanced uptake of levodopa into glial cells in the presence of COMT inhibition (Blessing et al., 2003). However, a reduction/increase of metabolites with potential toxic/protective properties of both levodopa and/or the methyl group metabolism induced by COMT inhibition (Fig. 1; Goetz, 1998; Männistö and Tuomainen, 1991; Storch et al., 2000a; Wurtman et al., 1972; Yassin et al., 1998) might also be involved in changing levodopa toxicity by COMT inhibition. Although changes of the levels of all major metabolites of levodopa/dopamine as well as SAM were excluded as a reason for altered levodopa toxicity by COMT inhibition (Blessing et al., 2003; Storch et al., 2000a), systematic data on dopaminergic toxicity induced *in vitro* by other products of the methyl group metabolism, such as HC and its derivatives, are not available. However, *in vitro* studies showed toxic effects of HC as well as its oxidation product homocysteic acid (HCA) on non-neuronal and neuronal cells (Althausen et al., 2000; Chen et al., 2000; Duan et al., 2002; Ho et al., 2002; Kim et al., 1996; Lipton et al., 1997; Parsons et al., 1998). Furthermore, HC potentiates the effects of several types of toxic insults to neuronal cultures, such as β-amyloid or copper (Ho et al., 2001; White et al., 2001). The neurotoxic mechanisms of HC are not fully understood, but stimulation of the *N*-methyl-D-aspartate (NMDA) receptor (Lipton et al., 1997), glutamate excitotoxicity, kinase hyperactivation (Ho et al., 2002), DNA damage (Ho et al., 2002), and induction of oxidative stress (Ho et al., 2001; Jara-Prado et al., 2003; White et al., 2001) seem to play major roles in HC toxicity.

Objective of the present study therefore was to investigate the toxic effects of HC and its derivatives SAH, HCA and cysteic acid (CA) on dopaminergic neurons in primary mesencephalic cultures from rat. Furthermore, we analyzed the effects of HC on dopaminergic neurons in mesencephalic cultures treated with the dopaminergic neurotoxin 1-methyl-4-phenylpyridinium (MPP^+) to simulate selective dopaminergic degeneration *in vitro*.

Materials and methods

Materials

Homocysteine (HC), *S*-adenosylhomocysteine (SAH), homocysteic acid (HCA), and cysteic acid (CA) were purchased from Sigma (St. Louis, MO). 1-Methyl-4-pyridinium iodide (MPP^+) was from Research Biochemical International (Natick, MA). All other chemicals were of analytical grade.

Primary mesencephalic cultures

Primary cultures from rat rostral mesencephalic tegmentum were prepared as previously described (Storch et al., 2000a,b). Briefly, the rostral mesencephalic tegmentum from E14.5 rat embryos (Charles-River, Braunschweig, Germany) was dissociated using trypsin, DNase and mechanical trituration. The cells were plated out at 125,000 viable cells/cm^2 on poly-L-lysine coated 48 well-plates in complete media containing DMEM (high glucose)/F-12 mixture (1:1), 1% penicillin/streptomycin (Sigma), and 10% fetal calf serum (PAA, Cölbe, Germany). Cultures were incubated at 37°C in a humidified atmosphere of 5% CO_2 in air. The culture medium was changed 24 h after plating.

Treatment paradigms

Primary cultures were treated after 4 days in culture with various concentrations of HC, SAH, HCA, and CA for 24 h or 48 h. To simulate dopaminergic degeneration *in vitro*, in some experiments the cells were pre-incubated with the selective dopaminergic neurotoxin MPP^+ (3, 10 or 30 μM) for

24 h prior to addition of various concentrations of HC (10 or 100 μM) for additional 24 h or the cultures were co-incubated with MPP^+ and HC for 48 h. Then, the cultures were further processed for immunocytochemical studies.

Quantification of dopaminergic cell survival in primary cultures

Following treatment, the cultures were fixed using 3.7% paraformaldehyde and then immunocytochemically processed for tyrosine hydroxylase (TH) using a monoclonal antibody against rat TH (1:1000; Sigma) and nickel enhancement (Storch et al., 2000a, b). The numbers of TH-positive cells were assessed by an investigator blinded to treatment history using a Zeiss inverted microscope, counting black cell bodies of the whole well surface area ($1\,cm^2$). Based on the tissue dissection procedure and the absence of immunoreactivity for the noradrenergic neuron marker dopamine-β hydroxylase in sister cultures, it is likely that TH-positive cells in our cultures were phenotypically dopaminergic neurons. In addition, we measured the number and length of cellular extensions (neurites) of TH-positive cells on representative digital microphotographs using the program VisRoute (Version 1.0.34 beta). Concerning the results of Koutsilieri and co-workers (1995) showing similar effects of the neurotoxin MPP^+ on morphology of both bipolar and multipolar TH-positive neurons in mesencephalic cultures, we combined the results on number and length of extensions of all TH-positive neurons (Koutsilieri et al., 1995). In a typical experiment, we measured the lengths of all extensions of a total of 50 TH-positive neurons per condition.

Statistics

To evaluate the toxic potency of MPP^+, 8 different concentrations (0 to 300 μM) were tested and the experiment was replicated six times. We calculated the half-maximal toxic concentration (TC_{50} value) by non-linear curve fitting (Origin, Version 5.0; MicroCal Software) according to the mathematical model $Y = A_1 + (A_2 - A_1)/(1 + 10^{(\log B - x)nH})$, where A_1 is the limit when the concentration approaches 0, A_2 the limit when the concentration approaches the maximum, B the effective concentration which leads to 50% reduction of cell viability (TC_{50}), and n_H the Hill slope. Statistical comparisons were made by Dunnett's *t*-test. If data were not normally distributed, a non-parametric test (Mann-Whitney U-test) was used for comparisons of results. All data are expressed as mean ± s.e.m.

Results

Effects of homocysteine and its derivatives on survival of dopaminergic neurons in vitro

Incubation of mesencephalic cultures with HC did not show significant effects on cell number of TH-positive neurons when administered alone at concentrations

Fig. 2. Effects of homocysteine and its derivatives on morphology of dopaminergic neurons in mesencephalic cultures. **A** Representative microphotographs demonstrate the effects of homocysteine (HC) on dopaminergic neurons after 48 h in absence and presence of MPP^+. (**a**) TH-positive neurons in a control culture at day 6 *in vitro*. TH-positive neurons in sister cultures exposed to (**b**) 10 μM MPP^+, (**c**) 100 μM HC, (**d**) 10 μM MPP^++100 μM HC. Bar represents 50 μm. **B** Effects of HC (100 μM) in absence and presence of MPP^+ (10 μM) as well as its derivatives homocysteic acid (HCA; 100 μM) and cysteic acid (CA; 100 μM) on length and number of neurites of dopaminergic neurons. After 4 days *in vitro* the cultures were incubated with the substances of interest for 48 h. Length and numbers of neurites were measured on digital microphotographs as described in the Material and method section. Data points represent the mean ± SEM of at least three independent experiments. * indicates $P < 0.05$; ** signifies $P < 0.01$ when compared to untreated control. Unpaired Student's *t*-test revealed no statistically significant differences in morphological parameters between cultures treated with HC and MPP^+ and cultures treated with MPP^+ alone

up to 1000 μM for 24 or 48 h, but led to changes on morphology of TH-positive neurons with significant fewer and shorter neurites (extensions) at 100 μM of HC after 48 h compared to control conditions (Figs. 2, 3A). As shown previously (Blessing et al., 2003), SAH showed significant toxic effects on TH-positive cells only at concentrations ≥1000 μM when administered for 24 h. After a 48 h exposure, SAH led to a significant TH-positive cell loss at

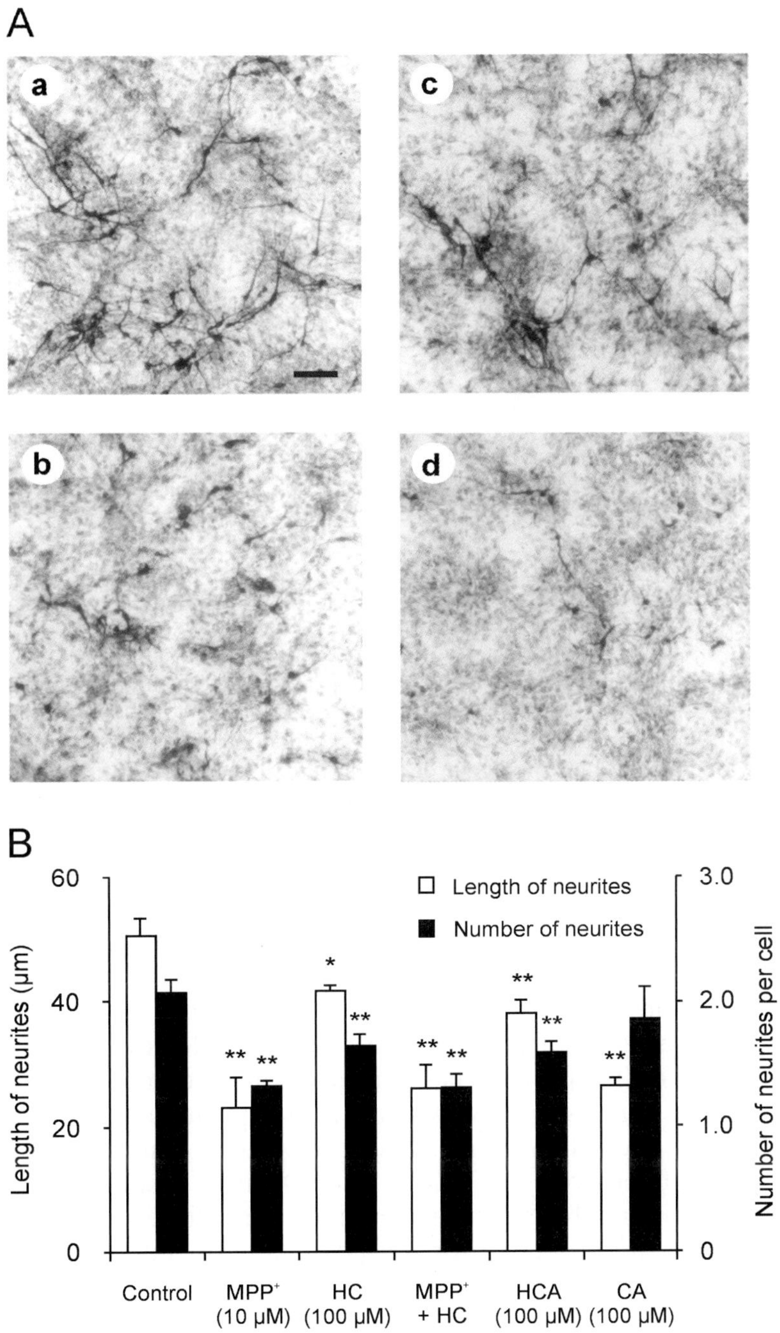

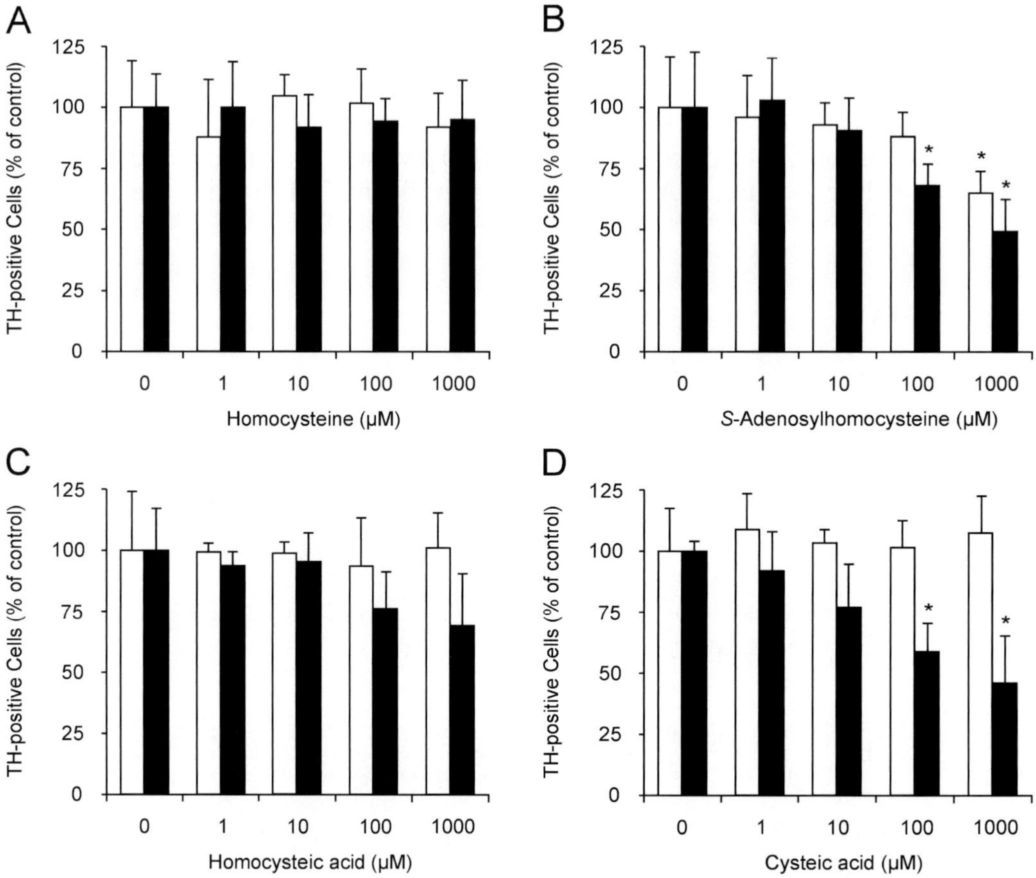

Fig. 3. Effects of homocysteine (**A**), *S*-adenosylhomocysteine (**B**), homocysteic acid (**C**) and cysteic acid (**D**) on survival of dopaminergic neurons after 24 h (open bars) and 48 h (closed bars). Survival of TH-positive neurons is expressed as percentage of untreated control. Control values were 382 ± 91 and 358 ± 54 TH-positive cells/well after 24 and 48 h, respectively. Data points represent the mean $\pm$ SEM of at least three independent experiments. * indicates $P < 0.05$ when compared to untreated control. Data on toxicity of SAH after a 24 h incubation period are from Blessing and co-workers (2003)

concentrations of $\geq 100\,\mu M$ with $68 \pm 9\%$ and $49 \pm 13\%$ TH-positive cell survival at 100 and $1000\,\mu M$, respectively ($P < 0.05$; Fig. 3B). Furthermore, the oxidation product of HC, homocysteic acid (HCA), did not show toxic effects on TH-positive neuron survival in mesencephalic cultures at concentrations up to $1000\,\mu M$ after exposure for up to 48 h (Fig. 3C). However, HCA ($100\,\mu M$) induced a change of TH-positive neuron morphology comparable to HC after 48 h (Fig. 2B). In contrast, CA was toxic to TH-positive neurons at concentration $\geq 100\,\mu M$ after 48 h (with TH-positive cell survival of $59 \pm 12\%$ and $46 \pm 19\%$ of control at 100 and $1000\,\mu M$, respectively; $P < 0.05$; Figs. 2B, 3D). Gross microscopic inspection of the cultures revealed that these effects of CA were not selective for TH-positive cells, but also influence cell count and morphology of all cell types in mesencephalic cultures (data not shown).

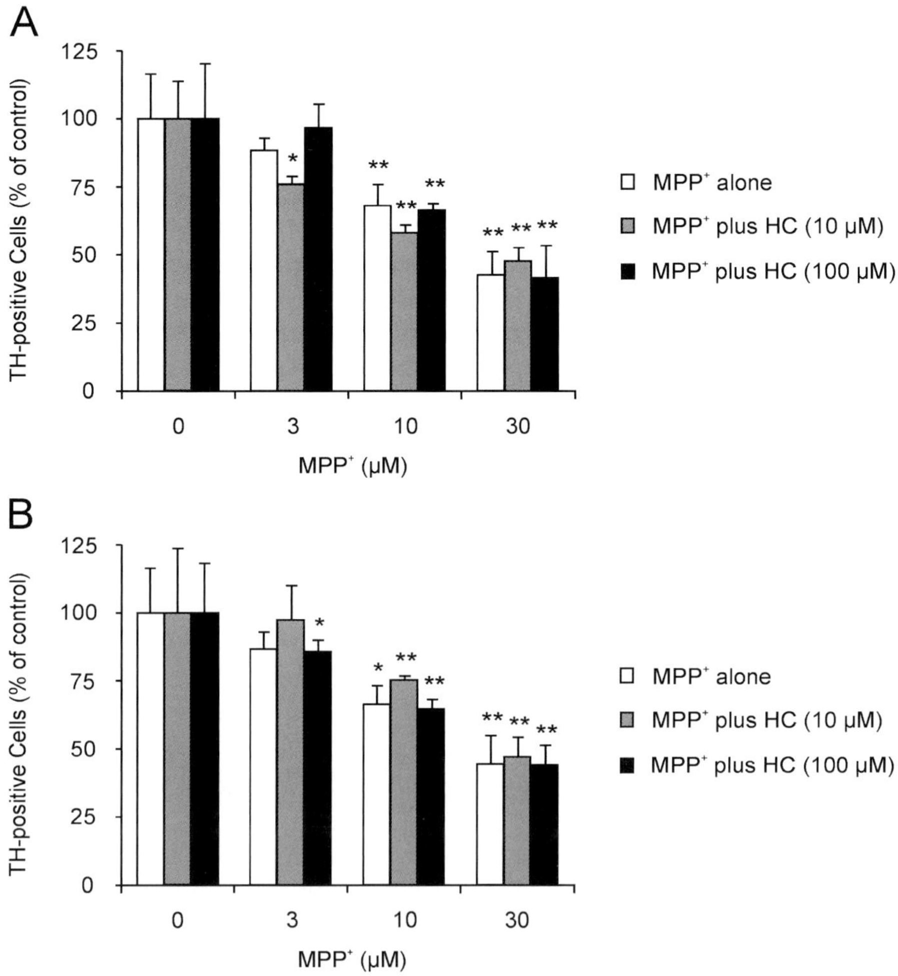

Fig. 4. Effects of homocysteine on dopaminergic neuron survival in mesencephalic cultures treated with 1-methyl-4-phenylpyridinium (MPP^+). **A** After 4 days *in vitro* the cultures were pre-incubated with MPP^+ (0, 3, 10 and 30 μM) for 24 h. Then, the cultures were exposed to various concentrations of homocysteine (0, 10 and 100 μM) in absence and presence of MPP^+, respectively, for additional 24 h. **B** After 4 days *in vitro* the cultures were co-incubated with various concentration of MPP^+ (0, 3, 10 and 30 μM) and homocysteine (0, 10 and 100 μM) for 48 h. In both conditions, survival of TH-positive neurons is expressed as percentage of untreated control. Data points represent the mean ± SEM of at least three independent experiments. Unpaired Student's *t*-test revealed no statistically difference in TH-positive neuron survival in cultures treated with various concentrations of homocysteine relative to control cultures (MPP^+ alone). * indicates $P<0.05$; ** signifies $P<0.01$ when compared to untreated control

Effects of homocysteine on dopaminergic neurons treated by MPP^+ in vitro

To investigate the effects of HC on degenerating dopaminergic neurons *in vitro*, we used the dopaminergic neurotoxin MPP^+ to selectively stress dopaminergic neurons in mesencephalic cultures. Prior to investigating the effects of HC, we determined the effects of MPP^+ on cell viability of dopaminergic neurons *in vitro*. MPP^+ caused a dose-related decrease of TH-positive cells in 4 day old mesencephalic cultures from rat, with a calculated TC_{50} value obtained from the toxicity curve by non-linear regression analysis of $19.7 \pm 11.4\,\mu M$ after 48 h of incubation ($n = 6$). As previously shown by Koutsilieri and co-workers (1995), the remaining TH-positive cells showed fewer and shorter neurites compared to cells without MPP^+ treatment (Fig. 2A, B), but gross microscopic inspection of the cultures revealed that MPP^+ at concentrations of $\leq 30\,\mu M$ did not influence morphology and cell count of other cell types. These observations are consistent with data reported in the literature (Michel et al., 1990). Using concentrations in the range of the TC_{50} value (3, 10 and 30 μM), the effects of various concentrations of HC on MPP^+-induced dopaminergic toxicity was determined (Figs. 2, 4). In cultures pre-incubated with MPP^+ for 24 h and then incubated with both MPP^+ and HC for additional 24 h, there were no further alterations of cell morphology and viability by HC compared to MPP^+ alone (Fig. 4A). In line with these results, a 48 h exposure of mesencephalic cultures to various concentrations of HC in the presence of MPP^+ did not lead to changes of TH-positive cell count and morphology compared to incubation with MPP^+ alone (Figs. 2, 4B).

Discussion

The metabolism of both levodopa and dopamine requires COMT-mediated *O*-methylation of the phenolic group of the catechol. Since COMT uses SAM as methyl donor, *O*-methylation of levodopa to 3-*O*-methyldopa as well as dopamine to 3-methyltyramine is associated with the conversion of SAM to SAH and subsequently to HC (Miller et al., 1997; Wurtman et al., 1972; Yassin et al., 1998). Consistently, elevated HC plasma levels appear in levodopa treated PD patients compared to matched controls and correlated to daily levodopa dosage (Allain et al., 1995; Kuhn et al., 1998; Yasui et al., 2000). Furthermore, we showed that treated but not untreated PD patients showed augmented plasma concentrations of HC (Müller et al., 1999). In addition, levodopa treatment in rats decreases the brain levels of SAM, whereas the levels of SAH and HC are elevated by levodopa treatment (Liu et al., 2000; Miller et al., 1997; Wurtman et al., 1972; Yassin et al., 1998). Consistently, we recently demonstrated these altered levels of substrates of *O*-methylation cycle in treated PD patients (Müller et al., 2001). HC induces substantial impairment of endothelial function with subsequent arteriosclerosis and occurrence of cardiovascular disease and stroke, both of which are reported to be increased in treated PD patients (Gorell et al., 1994; Ben Shlomo et al., 1995; Chambers et al., 1998). Furthermore, *in vitro* studies revealed toxic properties of HC and its derivatives HCA and CA on various cell types, including cultured human neuronal cell lines and primary cortical neurons from rat (Buemi et al., 2001; Chen et al., 2000; Lipton et al., 1997; Parsons et al., 1998). The exact mechanisms of toxicity are unclear, but excitotoxic actions of HC via *N*-methyl-D-aspartate (NMDA) receptor stimulation might be involved (Lipton et al., 1997). Duan and co-workers (2002) showed that HC sensitizes both dopaminergic cells *in vivo* and the neuronal cell line SK-N-MC *in vitro* to apoptosis under conditions of mitochondrial impairment and oxidative stress (Duan et al., 2002). These data led to the hypothesis that HC might be involved in the pathogenesis of PD and/or in enhanced progression of PD in patients treated with levodopa.

In the present study we investigated the toxicity of HC and its derivatives SAH, HCA and CA on dopaminergic neurons in primary mesencephalic cultures from rat *in vitro*. We used a wide range of HC concentrations (1 to 1000 μM), which reflects the clinical spectrum of plasma HC concentrations in PD patients (Müller et al., 1999, 2001). Indeed, plasma levels of HC in levodopa treated PD patients are between 8 and 25 μM (Müller et al., 1999, 2001). HC concentrations in normal mammalian brain tissue were reported to be up to 10 μM (Broch et al., 1984). HC and HCA did not show significant effects on dopaminergic cell viability with concentrations up to 1 mM, but led to morphological changes of dopaminergic neurons at a concentration of 100 μM as an expression of very mild toxicity of these compounds towards dopaminergic neurons. In contrast, SAH and CA displayed relevant toxicity at concentrations of >100 μM. These results are in line with data on human neuronal SK-N-MC cells with partial dopaminergic characteristics (Duan et al., 1999) demonstrating no toxicity of HC *in vitro* (Duan et al., 2002). Parsons and co-workers (1998) showed that HC is selectively toxic to medulloblastoma cells, but not to other neuronal cell types *in vitro* (Parsons et al., 1998). Using primary neuronal cortical or cerebellar granule cell cultures, several *in vitro* studies showed toxic effects of HC in the high micromolar concentration range (>200 μM) (Ho et al., 2002; Kim et al., 1996; Lipton et al., 1997). In addition, systematic investigations *in vivo* demonstrated that HC when administered alone does not alter dopaminergic cell count in the *substantia nigra pars compacta* of C57Bl/6 mice (Duan et al., 2002). The neurotoxic mechanisms of HC most likely include stimulation of the *N*-methyl-D-aspartate (NMDA) receptor (Lipton et al., 1997), glutamate toxicity, kinase hyperactivation (Ho et al., 2002), DNA damage (Ho et al., 2002), and induction of oxidative stress (Ho et al., 2001; Jara-Prado et al., 2003; White et al., 2001). The differences of the involvement of glutamate toxicity in response to toxic insults between cortical/cerebellar and mesencephalic cultures with no major role of glutamate toxicity in dopaminergic cultures (Michel and Agid, 1992) might explain the discrepancies of neuronal sensitivity against HC toxicity.

HCA does not display relevant toxicity towards all glial and neuronal cell lines tested by Parson and co-workers (1998). On the other hand, in accordance to the present study CA displays weak toxicity to various neuronal and glial cell types (Parson et al., 1998). The mechanisms of CA toxicity are unknown, but it seems unlikely that CA display selective neuronal/glial toxicity (Parson et al., 1998). The discrepancy that HC display no relevant dopaminergic toxicity whereas its metabolite CA showed weak toxicity in our *in vitro* system is most likely due to an insufficient catabolism of HC into CA in our cell culture system to reach CA concentrations leading to relevant dopaminergic toxicity.

We next tested the effects of HC on dopaminergic cell survival in mesencephalic cultures stressed with the dopaminergic neurotoxin and mitochondrial complex I inhibitor MPP^+. Using this neurotoxin, we simulated in our cell cultures system the selective dopaminergic degeneration found in PD (Lang and Lozano, 1998). Whereas MPP^+ showed strong toxicity towards dopaminergic neurons, co-incubation with HC did not further exacerbate dopaminergic cell death induced by MPP^+. These results are in contrast to data of Duan and co-workers (2002) demonstrating that HC enhances the toxicity of both mitochondrial inhibition by rotenone and oxidative stress by ferrous iron in SK-N-MC cells. Furthermore, HC potentiates the effects of several other types of toxic insults to primary cortical cultures, such as β-amyloid or copper (Ho et al., 2001; White et al., 2001). Interestingly, administration of HC together with 1-methyl-4-phenyl-1,2,3,6-tetrahydopyridine (MPTP), the blood-brain barrier trespassing precursor of MPP^+, to C57Bl/6 mice resulted in significant exacerbation of MPTP toxicity towards dopaminergic neurons *in vivo* (Duan et al., 2002). The reported discrepancies between the present study using primary dopaminergic neurons *in vitro* and other experimental systems with respect to their sensitivity towards HC-mediated potentiation of toxic insults might be due to different methods and/or cell systems used. Significant differences between the experimental protocols include for example the use of serum in our study, whereas Duan and co-workers (2002) used serum-free conditions known to dramatically increase oxidative stress (Menzies et al., 2002). Another reason for the high sensitivity of neuronal cells towards HC-induced exacerbation of toxic insults is most likely the great importance of indirect toxic mechanisms, in particular glutamate excitotoxicity, in primary cortical/cerebellar cell cultures and the MPTP animal model (Klockgether et al., 1991; Loschmann et al., 1994; Turski et al., 1991). In contrast, in primary dopaminergic cultures like those used in the present study, major toxic mechanisms of HC, such as glutamate excitotoxicity and NMDA receptor agonism (Lipton et al., 1997; Ho et al., 2002), only play a minor role in the response to toxic insults, such as MPP^+ treatment (Michel and Agid, 1992).

Taken together, our results do not demonstrate relevant dopaminergic toxicity of HC and its major metabolites in primary mesencephalic cultures *in vitro*. Thus, production of SAH and subsequently HC/HCA during the conversion of levodopa to 3-*O*-methyldopa by the enzyme COMT is not responsible for the toxicity of levodopa *in vitro*. Consistently, reduced production of these compounds in the presence of COMT inhibition is not responsible for the protective effects of COMT inhibitors against levodopa (Storch et al., 2000; Blessing et al., 2003). Furthermore, HC does not exacerbate dopaminergic toxicity of the selective dopaminergic mitochondrial complex I inhibitor MPP^+ *in vitro*. Our data do not fully rule out that indirect actions of HC, such as excitotoxic effects (Lipton et al., 1997), play a relevant role in exacerbating MPTP toxicity by HC *in vivo*. However, the present data do not support the hypothesis that HC or its major derivatives are involved in the pathogenesis of PD or in accelerating the progression of PD in patients treated with levodopa.

References

Allain P, Le Bouil A, Cardillet E, Le Quay L, Bagheri H, Montastruc JL (1995) Sulfate and cysteine levels in the plasma of patients with Parkinson's disease. Neurotoxicology 16: 527–529

Althausen S, Paschen W (2000) Homocysteine-induced changes in mRNA levels of genes coding for cytoplasmic- and endoplasmic reticulum-resident stress proteins in neuronal cell cultures. Brain Res Mol Brain Res 84(1–2): 32–40

Ben Shlomo Y, Marmot MG (1995) Survival and cause of death in a cohort of patients with parkinsonism: possible clues to aetiology? J Neurol Neurosurg Psychiatry 58: 293–299

Blessing H, Bareiss M, Zettlmeisl H, Schwarz J, Storch A (2003) Catechol-*O*-methyltransferase inhibition protects against 3,4-dihydroxyphenylalanine (dopa) toxicity in primary mesencephalic cultures: new insights into levodopa toxicity. Neurochem Int 42(2): 139–151

Blunt SB, Jenner P, Marsden CD (1993) Suppressive effects of L-dopa on dopamine cells remaining in the ventral tegmental area of rats previously exposed to the neurotoxin 6-hydroxydopamine. Mov Disord 8: 129–133

Broch OJ, Ueland PM (1984) Regional distribution of homocysteine in the mammalian brain. J Neurochem 43(6): 1755–1757

Buemi M, Marino D, Di Pasquale G, Floccari F, Ruello A, Aloisi C, Corica F, Senatore M, Romeo A, Frisina N (2001) Effects of homocysteine on proliferation, necoris and apoptosis of vascular smooth muscle cells in culture and influence of folic acid. Thromb Res 104: 207–213

Chambers JC, McGregor A, Kooners JS (1998) Acute hyperhomocysteinaemia and endothelial dysfanction. Lancet 351: 36–37

Chen C, Halkos ME, Surowiec SM, Conklin BS, Lin PH, Lumsden AB (2000) Effects of homocysteine on smooth muscle cell proliferation in both cell culture and artery perfusion culture models. J Surg Res 88: 26–33

Duan W, Zhang Z, Gash DM, Mattson MP (1999) Participation of prostate apoptosis response-4 in degeneration of dopaminergic neurons in models of Parkinson's disease. Ann Neurol 46: 587–597

Duan W, Ladenheim B, Cutler RG, Kruman II, Cadet JL, Mattson MP (2002) Dietary folate deficiency and elevated homocysteine levels endanger dopaminergic neurons in models of Parkinson's disease. J Neurochem 80: 101–110

Fahn S, Parkinson's Study Group (2002) Results of the ELLDOPA (Earlier vs. Later levodopa) study. Mov Disord 17 [Suppl 5]: S13

Goetz CG (1998) Influence of COMT inhibition on levodopa pharmacology and therapy. Neurology 50 [Suppl 5]: S26–S30

Gorell JM, Johnson CC, Rybicki BA (1994) Parkinson's disease and its comorbid disorders: an analysis of Michigan martality data, 1970 to 1990. Neurology 44: 1865–1868

Ho PI, Collins SC, Dhitavat S, Ortiz D, Ashline D, Rogers E, Shea TB (2001) Homocysteine potentiates beta-amyloid neurotoxicity: role of oxidative stress. J Neurochem 78(2): 249–253

Ho PI, Ortiz D, Rogers E, Shea TB (2002) Multiple aspects of homocysteine neurotoxicity: glutamate excitotoxicity, kinase hyperactivation and DNA damage. J Neurosci Res 70(5): 694–702

Jara-Prado A, Ortega-Vazquez A, Martinez-Ruano L, Rios C, Santamaria A (2003) Homocysteine-induced brain lipid peroxidation: effects of NMDA receptor blockade, antioxidant treatment, and nitric oxide synthase inhibition. Neurotox Res 5(4): 237–243

Jeffrey D, Roth JA (1984) Characterization of membrane-bound and soluble catechol-*O*-methyltransferase from human frontal cortex. J Neurochem 24(3): 826–832

Kim WK, Pae YS (1996) Involvement of N-methyl-D-aspartate receptor and free radical in homocysteine-mediated toxicity on rat cerebellar granule cells in culture. Neurosci Lett 216(2): 117–120

Klockgether T, Turski L, Honore T, Zhang ZM, Gash DM, Kurlan R, Greenamyre JT (1991) The AMPA receptor antagonist NBQX has antiparkinsonian effects in monoamine depleted rats and MPTP-treated monkeys. Ann Neurol 30: 717–723

Koutsilieri E, Chen T-S, Kruzik P, Rausch W-D (1995) A morphometric analysis of bipolar and multipolar TH-IR neurons treated with the neurotoxin MPP^+ in co-cultures from mesencephalon and striatum of embryonic C57BL/6 mice. J Neurosci Res 41: 197–205

Kruman II, Kumaravel TS, Lohani A, Pedersen WA, Cutler RG, Kruman Y, Haughey N, Lee J, Evans M, Mattson MP (2002) Folic acid deficiency and homocysteine impair DNA repair in hippocampal neurons and sensitize them to amyloid toxicity in experimental models of Alzheimer's disease. J Neurosci 22(5): 1752–1762

Kuhn W, Roebroek R, Blom H, van Oppenraaij D, Müller T (1998) Hyperhomocysteinaemia in Parkinson's disease. J Neurol 245: 811–812

Lang AE, Lozano AM (1998) Parkinson's disease. N Engl J Med 339: 1044–1053; 1130–1143

Lindgren A, Brattstrom L, Norrving B, Hultberg B, Andersson A, Johansson BB (1995) Plasma homocysteine in the acute and convalescent phases after stroke. Stroke 26(5): 795–800

Lipton SA, Kim WK, Choi YB, Kumar S, D'Emilia DM, Rayudu PV, Arnelle DR, Stamler JS (1997) Neurotoxicity associated with dual actions of homocystein at the N-methyl-D-aspartate receptor. Proc Natl Acad Sci USA 94: 5923–5928

Liu XX, Wilson K, Charlton CG (2000) Effects of L-Dopa treatment on methylation in mouse brain: implications for the side effects of L-dopa. Life Sci 66: 2277–2288

Loschmann PA, Lange KW, Wachtel H, Turski L (1994) MPTP-induced degeneration: interference with glutamatergic toxicity. J Neural Transm [Suppl] 43: 133–143

Luquin R, Del Rio L, Saldise L, Setuain J, Manzi B (2001) Levodopa but not pergolide administration induces apoptosis in the substantia nigra neurons of rats with partial nigrostriatal lesion. Neurology 56 [Suppl 3]: A378

Mannistö PT, Tuomainen P (1991) Effect of high single doses of levodopa and carbidopa on brain dopamine and its metabolites: modulation by selective inhibitors of monoamine oxidase and/or catechol-*O*-methyltransferase in the male rat. Naunyn Schmiedebergs Arch Pharmacol 344(4): 412–418

Männistö PT, Ulmanen I, Lundström K, Taskinen J, Tenhunen J, Tilgmann C, Kaakkola S (1992) Characteristics of catechol *O*-methyl transferase (COMT) and properties of selective COMT inhibitors. Prog Drug Res 39: 291–350

Mena MA, Pardo B, Casarejos MJ, Fahn S, De Yebenes JG (1992) Neurotoxicity of levodopa on catecholamine-rich neurons. Mov Disord 7: 23–31

Mena MA, Pardo B, Paino CL, De Yebenes JG (1993) Levodopa toxicity in foetal rat midbrain neurones in culture: modulation by ascorbic acid. NeuroReport 4(4): 438–440

Menzies FM, Cookson MR, Taylor RW, Turnbull DM, Chrzanowska-Lightowlers ZM, Dong L, Figlewicz DA, Shaw PJ (2002) Mitochondrial dysfunction in a cell model of familial amyotrophic lateral sclerosis. Brain 125: 1522–1533

Michel PP, Agid Y (1992) The glutamate antagonist, MK-801, does not prevent dopaminergic cell death induced by the 1-methyl-4-phenylpyridinium ion (MPP^+) in rat dissociated mesencephalic cultures. Brain Res 597: 233–240

Michel PP, Dandapani BK, Knusel B, Sanchez-Ramos J, Hefti F (1990) Toxicity of 1-methyl-4-phenylpyridinium for rat dopaminergic neurons in culture: selectivity and irreversibility. J Neurochem 54: 1102–1109

Miller JW, Shukitt-Hale B, Villalobos-Molina R, Nadeau MR, Selhub J, Joseph JA (1997) Effect of L-Dopa and the catechol-*O*-methyltrasnferase inhibitor Ro 41-0960 on sulfur amino acid metabolites in rats. Clin Neuropharmacol 20: 55–66

Müller T, Werne B, Fowler B, Kuhn W (1999) Nigral endothelial dysfunction, homocystein, and Parkinson's disease. Lancet 354: 126–127

Müller T, Woitalla D, Hauptmann B, Fowler B, Kuhn W (2001) Decrease of methionine and *S*-adenosylmethionine and increase of homocysteine in treated patients with Parkinson's disease. Neurosci Lett 308: 54–56

Parsons RB, Waring RH, Ramsden DB, Williams AC (1998) *In vitro* effects of cysteine metabolites homocysteic acid, homocysteine and cysteic acid upon human neuronal cell lines. Neurotoxicology 19: 599–603

Perry IJ, Refsum H, Morris RW, Ebrahim SB, Ueland PM, Shaper AG (1995) Prospective study of serum total homocysteine concentration and risk of stroke in middle-aged British men. Lancet 346: 1395–1398

Selhub J, Jacques PF, Bostom AG, D'Agostino RB, Wilson PW, Belanger AJ, O'Leary DH, Wolf PA, Schaefer EJ, Rosenberg IH (1995) Association between plasma homocysteine concentrations and extracranial carotid-artery stenosis. N Engl J Med 332(5): 286–291

Storch A, Blessing H, Bareiss M, Jankowski S, Ling ZD, Carvey P, Schwarz J (2000a) Catechol-*O*-methyltransferase inhibition attenuates levodopa toxicity in mesencephalic dopamine neurons. Mol Pharmacol 57: 589–594

Storch A, Burkhardt K, Ludolph AC, Schwarz J (2000) Protective effects of riluzole on dopamine neurons: involvement of oxidative stress and cellular energy metabolism. J Neurochem 75(6): 2259–2269

Tolosa E, Marti MJ, Valldeoriola F, Molinuevo JL (1998) History of levodopa and dopamine agonists in Parkinson's disease treatment. Neurology 50 [Suppl 6]: S2–S10

Turski L, Bressler K, Rettig KJ, Löschmann PA (1991) Protection of substantia nigra from MPP^+ neurotoxicity by *N*-methyl-D-aspartate antagonists. Nature 349: 414–418

White AR, Huang X, Jobling MF, Barrow CJ, Beyreuther K, Masters CL, Bush AI, Cappai R (2001) Homocysteine potentiates copper- and amyloid beta peptide-mediated toxicity in primary neuronal cultures: possible risk factors in the Alzheimer's-type neurodegenerative pathways. J Neurochem 76(5): 1509–1520

Wurtman RJ (1972) Effect of L-dopa on *S*-adenosylmethionine levels and norepinephrine metabolism in rat brain. In: Costa E, Iversen LL, Paoletti R (eds) Advances in biochemical psychopharmacology, vol 6. Raven Press, New York, pp 241–246

Yassin MS, Cheng H, Ekblom J, Oreland L (1998) Inhibitors of catecholamine metabolizing enzymes cause changes in *S*-adenosylmethionine and *S*-adenosylhomocystein in the rat brain. Neurochem Int 32: 53–59

Yasui K, Kowa H, Nakaso K, Takeshima T, Nakashima K (2000) Plasma homocysteine and MTHFR C677T genotype in levodopa-treated patients with PD. Neurology 55: 437–440

Authors' address: A. Storch, M.D., Department of Neurology, Technical University of Dresden, Fetscherstrasse 74, D-01307 Dresden, Germany, e-mail: alexander.storch@neuro.med.tu-dresden.de

J Neural Transm (2004) [Suppl] 68: 15–20

MTHFR C677T polymorphism, folic acid and hyperhomocysteinemia in levodopa treated patients with Parkinson's disease

D. Woitalla, W. Kuhn, and **T. Müller**

Department of Neurology, St. Josef Hospital, Ruhr University, Bochum, Germany

Summary. Certain mutations (*TT* homozygous; *CT* heterozygous; *CC* wild-type) of the methylenetetrahydrofolate (MTHFR) gene and long-term levodopa application in patients with Parkinson's disease (PD) support onset of hyperhomocysteinemia. Total plasma homocysteine (t-hcys) depends on B_6, B_{12}, folic acid, all of which support remyelination from t-hcys to methionine. Objective of this trial were to compare B_6, B_{12}, folic acid and t-hcys levels in plasma of 83 levodopa treated PD patients and 44 controls. PD patients with the CT or TT genotype had significant higher t-hcys levels than controls or PD patients with the CC allele. Concentrations of B_6 or B_{12} did not differ, but folic acid was significant higher in PD patients with the CT mutation. We recommend MTHFR genotyping, t-hcys monitoring and early vitamin supplementation in PD patients. The folic acid increase in PD patients with the CT allele is hypothetically due to an endogenous upregulation of folic acid absorption to decrease t-hcys.

Introduction

Levodopa is administered with dopa decarboxylase inhibitors (DDI) to prevent its peripheral degradation. This increases conversion of levodopa to 3-O-methyldopa (3-OMD) by the ubiquitious enzyme catechol-O-methyltransferase (COMT) in blood, peripheral tissues and in nigrostriatal neurons. COMT requires Mg^{2+} as cofactor and S-adenosylmethionine as methyl donor. Thus O-methylation of levodopa to 3-OMD is associated with conversion of S-adenosylmethionine (SAM) to S-adenosylhomocysteine (SAH) and subsequently homocysteine (Müller et al., 2001, 2002). Therefore long-term levodopa/DDI metabolism via COMT results in elevated total homocysteine (t-hcys)-, decreased methionine- and SAM plasma levels in patients with Parkinson's Disease (PD) (Müller et al., 2001). However, these putative neurotoxic long-term levodopa/DDI effects due to altered homocysteine-, SAM-, and SAH metabolism associated are not clinically proven in treated PD patients. But increased t-hcys levels contribute to incidence of vascular disease, brain atrophy, Alzheimer's disease or all of them (Dekou et al., 2001; Sachdev et al., 2002; Seshadri and Wolf, 2003; Bleich and Kornhuber, 2003; Bleich et al., 2003). A further genetic predisposing factor for t-hcys concentrations is a polymorphism in the gene encoding the thermolabile enzyme methylenetetrahydrofolate reductase (MTHFR) (Ueland et al., 2001). MTHFR catalyzes the remethylation of homocysteine to methionine in the presence of folic acid, vitamins B_6 and B_{12}. Homozygosity for the C-to-T substitution at nucleotide 677 of the MTHFR-gene leads to a 50% reduction of the activity of this enzyme and is the most common inherited cause of moderate hyperhomocysteinemia (Ueland et al., 2001). However, there are only rare detailed data on vitamin – and t-hcys metabolism and their relation to each other and the MTHFR polymorphism in levodopa/DDI treated PD patients (Miller et al., 2003). Objective of this trial were to compare levels of B_6, B_{12}, folic acid and t-hcys in plasma of PD patients, subdivided according to their MTHFR C677T genotypes, and of controls.

Subjects and methods

We determined MTHFR C677T genotypes (*TT* homozygous; *CT* heterozygous; *CC* wild-type) in 83 levodopa-treated PD patients (clinical characteristics; Table 1). The MTHFR C to T677 substitution was identified with the use of restriction enzyme digestion of the polymerase chain reaction-amplified products (Yasui et al., 2000; Dekou et al., 2001; Ueland et al., 2001). Prior enrollment, we excluded individuals with metabolic disturbances, i.e. diabetes, abnormal vitamin values or performed vitamin supplementation. We did not include subjects with neurological diseases other than PD. Each PD patient and each control subject fasted and was off any drug therapy for at least 12 hours before taking of blood samples, which were then centrifuged, decanted and stored at −80°C. The time period between freezing and work up of the plasma samples was no longer than three months. We measured levels of t-hys by an automated high performance liquid chromatography (HPLC) with reverse phase separation and fluorescent detection in all study participants (Ueland et al., 1993). We measured plasma concentrations of B_6, B_{12}, folic acid in all PD patients and 44 controls (clinical characteristics: see Table 1) with a conventional assay, which is established in our clinical routine according to previous trials (Müller et al., 2003).

Ethics

Each subject gave informed consent. The local ethics committee of the university approved this study.

Statistics

We used analysis of covariance (ANCOVA) and set age and sex as covariates for comparisons. We employed Tukeys Honest Significance Difference Test for different numbers for the post hoc analysis. We used linear regression for correlation analysis. The level of significance was $p < 0.05$.

Results

There were significant different t-hcys concentrations in PD patients and controls. We found in the post hoc analysis, that PD patients with the CT and TT genotype had higher t-hcys levels than controls, respectively PD patients with the CC mutation (Tables 1, 2). Concentration of folic acid also significantly differed in the investigated groups. PD patients with the CT allele had higher

Table 1. Clinical characteristics

	PD	CT	TT	CC	CO
Number	83	38	12	33	44
Age	65.28 ± 10.29; 41–88	63.11 ± 9.16; 48–88	66.33 ± 11.52; 41–81	67.39 ± 10.87; 48–84	58.14 ± 12.42; 24–80
Sex	47 men, 36 women	19 men, 19 women	9 men, 3 women	19 men, 14 women	21 men, 20 women
LD/DDI [mg]	440.81 ± 239.27; 100–1075	427.30 ± 260.29; 100–1075	450 ± 207.8; 150–900	453.03 ± 230.47; 100–900	
Duration of PD	6.54 ± 5.64; 0.1–22	7.49 ± 6.17; 0.1–22	6.13 ± 5.03; 0.1–15	5.61 ± 5.18; 0.1–20	
HYS	2.89 ± 0.94; I–IV	2.84 ± 0.95; I–V	2.67 ± 1.15; I–V	3.03 ± 0.85; I–V	
t-hcys [μmol/l]	17.9 ± 9.84; 4.9–72	14 ± 5.7; 4.9–31.5	22 ± 8.58; 11–38.50	20.9 ± 12.3; 8.4–72	13.01 ± 5.49; 5.6–30
B_6 [mg/l]	18.7 ± 6.69; 5.2–39.5	20.65 ± 7.49; 6.5–39.5	18.25 ± 5.72; 11–33.5	16.62 ± 5.43; 5.2–32.5	19.23 ± 8.71; 10–63.5
B_{12} [mg/l]	371.45 ± 218.43; 139–1611	322.55 ± 119.89; 139–699	369.58 ± 172.5; 169–693	428.42 ± 298.7; 166–1611	417.86 ± 232.95; 215–1389
Folic acid [mg/l]	6.39 ± 3.31; 2.4–21	7.53 ± 3.95; 2.50–21.00	5.54 ± 2.79; 2.5–12.6	5.39 ± 2.14; 2.4–13	6.48 ± 2.86; 3–15.7

Age and duration of PD (Parkinson's disease) are given in years, *LD/DDI* daily dosage of levodopa/dopa decarboxylase inhibitor intake in milligram [mg], *HYS* range of the Hoehn and Yahr Scale, *t-hcys* total plasma level of homocysteine; MTHFR C677T genotypes (*TT* homozygous; *CT* heterozygous; *CC* wild-type); *Co* Controls; *PD* patients with Parkinson's disease

Table 2. Results

	ANCOVA		Post hoc analysis (Tukeys HSD test)					
	F	p	CT vs TT	CT vs CC	CT vs CO	TT vs CC	TT vs CO	CC vs CO
Age	5.34	0.002	0.995	0.384	0.003	0.89	0.27	0.2
LD/DDI	0.11	0.9	0.99	0.9		0.97		
Duration of PD	1.02	0.37	0.97	0.37		0.83		
t-hcys	8.7	**5.33E − 05**	0.99	**0.0029**	**0.0004**	0.07	**0.03**	0.95
B_6	1.9	0.13	0.95	0.12	0.48	0.86	0.99	0.84
B_{12}	1.73	0.16	0.917	0.218	0.997	0.96	0.95	0.25
Folic acid	3.46	**0.019**	0.999	**0.025**	0.479	0.39	0.89	0.45

F F-value of ANCOVA; *p* p-value; *t-hcys* total plasma level of homocysteine; *Co* Controls; MTHFR C677T genotypes (*TT* homozygous; *CT* heterozygous; *CC* wild-type), significant results are bold

folic acid levels compared with the other ones according to the post hoc analysis (Tables 1, 2). No relevant changes of B_6 and B_{12} plasma concentrations occurred (Tables 1, 2). There was no impact of the covariates on our results.

The correlation analysis showed significant associations between B_6 and duration of PD ($R = 0.3$; $p = 0.005$), folic acid and t-hcys (PD patients and controls [N = 123], $R = -0.35$; $p = 6.58E-05$; PD patients [N = 83]; $R = -0.38$; $p = 0.0003$), and dosage of daily levodopa/DDI intake and t-hcys (PD patients [N = 83]; $R = 0.22$; $p = 0.04$). No further significant relations were found.

Discussion

We confirm previous results with significantly increased homocysteine levels in PD patients compared with controls and its relation to the daily levodopa/DDI intake (Kuhn et al., 1998). Yasui et al. (2000) found significantly elevated plasma homocysteine levels in 90 levodopa-treated PD patients with a clear but not significant increase of t-hcys levels in the homozygous MTHFR C677T genotype. In contrast, t-hcys concentrations were significantly higher both in PD patients with a heterozygous and homozygous mutation of the gene encoding MTHFR in our present study. Thus both levodopa/DDI treatment and polymorphism of MTHFR are risk factors for inducing moderate hyperhomocysteinemia and subsequently related vascular disease in PD patients (Gorell et al., 1994; Ben Shlomo and Marmot, 1995; Kuhn et al., 2001; Müller, 2002). We confirm the association between levels of folic acid and t-hcys in our cohort of PD patients (Miller et al., 2003). We hypothesize, that the significant moderate folic acid increase in PD patients with the CT allele may result from an endogenous upregulation of folic acid absorption in order to avoid t-hcys elevation and to balance altered methylation in PD patients (Williams et al., 1991, 1993).

Since one may assume, that levodopa-treated PD patients do not increase their intake of labile methyls, they must synthesize all their extra methyls consumed for levodopa methylation. If methyl neogenesis is hampered by poor folate intake, especially if the patient is a TT-homozygote for C677T/MTHFR polymorphism or has a low vitamin B_{12} level, production of SAM methyls may not be sufficient for both methylation of endogenous compounds and levodopa. This may lead to hypomethylation of essential neural components, and might explain the neuropsychiatric symptoms in folate and vitamin B_{12} deficiency (Brattstrom, 2001). This mechanism may contribute to side effects of levodopa. If this is the case, early combination of levodopa with COMT inhibitors would prove to be beneficial (Brattstrom, 2001; Müller, 2002; Müller et al., 2003). This warrants further studies on the relationship between complications to levodopa therapy in PD and homocysteine, folate, vitamin B_{12}, and C677T/MTHFR genotypes, like the report of a mild hypertrophy of the intima-media complex (IMC) of the carotid artery in long-term levodopa treated PD patients. The hypertrophic IMC, which is an established marker of systemic atherosclerosis, was not found in PD patients without prior levodopa treatment. In this study, PD patients with hypertrophic IMC of the carotid artery also exhibited elevated plasma levels of t-hcys associated with the C677T genotype of MTHFR. Moreover, a prolonged duration of treatment with levodopa in patients with the MTHFR T/T genotype enhanced the hypertrophy of IMC, compared with patients with the C/C or C/T genotype. These results suggest that hyperhomocysteinemia promoted by the C677T genotype of MTHFR and prolonged treatment with levodopa enhances atherosclerosis in PD patients and thus affects their general condition (Nakaso et al., 2003).

In conclusion, levodopa can cause hyperhomocysteinemia in PD patients, the extent of which is influenced by folic acid and the MTHFR mutation. Therefore we suggest MTHFR genotyping before initiation of levodopa/DDI long-term treatment. This would identify PD patients with an increased risk for levodopa/DDI associated t-hcys elevation. We recommend t-hcys monitoring particularly in PD patients with the TT or CT allele, in order to avoid putative chronic neurotoxicity via altered homocysteine metabolism. Moreover our results warrant early supplementation with folic acid or vitamins B_6 or B_{12} in these PD patients, since all of them support remethylation of homocysteine to methionine (Malinow et al., 1997; Miller et al., 2003).

References

Ben Shlomo Y, Marmot MG (1995) Survival and cause of death in a cohort of patients with parkinsonism: possible clues to aetiology? J Neurol Neurosurg Psychiatry 58: 293–299

Bleich S, Kornhuber J (2003) Relationship between plasma homocysteine levels and brain atrophy in healthy elderly individuals. Neurology 60: 1220

Bleich S, Bandelow B, Javaheripour K, Muller A, Degner D, Wilhelm J, Havemann-Reinecke U, Sperling W, Ruther E, Kornhuber J (2003) Hyperhomocysteinemia as a new risk factor for brain shrinkage in patients with alcoholism. Neurosci Lett 335: 179–182

Brattstrom L (2001) Plasma homocysteine and MTHFR C677T genotype in levodopa-treated patients with PD. Neurology 56: 281–282

Dekou V, Whincup P, Papacosta O, Ebrahim S, Lennon L, Ueland PM, Refsum H, Humphries SE, Gudnason V (2001) The effect of the C677T and A1298C polymorphisms in the methylenetetrahydrofolate reductase gene on homocysteine levels in elderly men and women from the British regional heart study. Atherosclerosis 154: 659–666

Gorell JM, Johnson CC, Rybicki BA (1994) Parkinson's disease and its comorbid disorders: an analysis of Michigan mortality data, 1970 to 1990. Neurology 44: 1865–1868

Kuhn W, Roebroek R, Blom H, van Oppenraaij D, Müller T (1998) Hyperhomocysteinemia in Parkinson's disease. J Neurol 245: 811–812

Kuhn W, Hummel T, Woitalla D, Müller T (2001) Plasma homocysteine and MTHFR C677T genotype in levodopa-treated patients with PD. Neurology 56: 281–282

Malinow MR, Nieto FJ, Kruger WD, Duell PB, Hess DL, Gluckman RA, Block PC, Holzgang CR, Anderson PH, Seltzer D, Upson B, Lin QR (1997) The effects of folic acid supplementation on plasma total homocysteine are modulated by multivitamin use and methylenetetrahydrofolate reductase genotypes. Arterioscler Thromb Vasc Biol 17: 1157–1162

Miller JW, Selhub J, Nadeau MR, Thomas CA, Feldman RG, Wolf PA (2003) Effect of L-dopa on plasma homocysteine in PD patients: relationship to B-vitamin status. Neurology 60: 1125–1129

Müller T (2002) Dopaminergic substitution in Parkinson's disease. Expert Opin Pharmacother 3: 1393–1403

Müller T, Woitalla D, Hauptmann B, Fowler B, Kuhn W (2001) Decrease of methionine and S-adenosylmethionine and increase of homocysteine in treated patients with Parkinson's disease. Neurosci Lett 308: 54–56

Müller T, Woitalla D, Fowler B, Kuhn W (2002) 3-OMD and homocysteine plasma levels in parkinsonian patients. J Neural Transm 109: 175–179

Müller T, Woitalla D, Kuhn W (2003) Benefit of folic acid supplementation in parkinsonian patients treated with levodopa. J Neurol Neurosurg Psychiatry 74: 549

Nakaso K, Yasui K, Kowa H, Kusumi M, Ueda K, Yoshimoto Y, Takeshima T, Sasaki K, Nakashima K (2003) Hypertrophy of IMC of carotid artery in Parkinson's disease is associated with L-DOPA, homocysteine, and MTHFR genotype. J Neurol Sci 207: 19–23

Sachdev PS, Valenzuela M, Wang XL, Looi JC, Brodaty H (2002) Relationship between plasma homocysteine levels and brain atrophy in healthy elderly individuals. Neurology 58: 1539–1541

Seshadri S, Wolf PA (2003) Homocysteine and the brain: vascular risk factor or neurotoxin? Lancet Neurol 2: 11

Ueland PM, Refsum H, Stabler SP, Malinow MR, Andersson A, Allen RH (1993) Total homocysteine in plasma or serum: methods and clinical applications. Clin Chem 39: 1764–1779

Ueland PM, Hustad S, Schneede J, Refsum H, Vollset SE (2001) Biological and clinical implications of the MTHFR C677T polymorphism. Trends Pharmacol Sci 22: 195–201

Williams A, Sturman S, Steventon G, Waring R (1991) Metabolic biomarkers of Parkinson's disease. Acta Neurol Scand [Suppl] 136: 19–23

Williams AC, Pall HS, Steventon GB, Green S, Buttrum S, Molloy H, Waring RH (1993) N-methylation of pyridines and Parkinson's disease. Adv Neurol 60: 194–196

Yasui K, Kowa H, Nakaso K, Takeshima T, Nakashima K (2000) Plasma homocysteine and MTHFR C677T genotype in levodopa-treated patients with PD. Neurology 55: 437–440

Authors' address: T. Müller, MD, Department of Neurology, St. Josef Hospital, Ruhr University Bochum, Gudrunstrasse 56, D-44791 Bochum, Germany, e-mail: thomas.mueller@ruhr-uni-bochum.de

J Neural Transm (2004) [Suppl] 68: 21–28

Screening for mutations in synaptotagmin XI in Parkinson's disease

A. S. Glass[1], **D. P. Huynh**[2], **Th. Franck**[1], **D. Woitalla**[3], **Th. Müller**[3], **S. M. Pulst**[2], **D. Berg**[1,5], **R. Krüger**[4], and **O. Riess**[1]

[1] Department of Medical Genetics, University of Tübingen, Tübingen, Germany
[2] Division of Neurology, Cedars-Sinai Medical Center, David Geffen School of Medicine at UCLA, Los Angeles, CA, USA
[3] Department of Neurology, Ruhr-University Bochum, St. Josef-Hospital, Bochum,
[4] Department of Neurology, and
[5] Hertie Institute for Brain Research, University of Tübingen, Germany

Summary. Parkinson's disease (PD) is characterized by selective degeneration of neurons in the substantia nigra and subsequent dysfunction of dopaminergic neurotransmission. Genes identified in familial forms of PD encode proteins that are linked to the ubiquitin-proteasome system indicating the pathogenic relevance of disturbed protein degradation in PD. Some of them, i.e. α-synuclein, parkin and synphilin-1, have been implicated in presynaptic neurotransmission based on their localization in synaptic vesicles. Synaptotagmin XI is linked to the pathogenesis of PD based on its identification as a substrate of the ubiquitin-E3-ligase parkin. Moreover synaptotagmin XI is involved in the maintainance of synaptic function and represents a component of Lewy bodies (LB) in brains of PD patients. Therefore, we performed a detailed mutation analysis of the synaptotagmin XI gene in a large sample of 393 familial and sporadic PD patients. We did not find any disease causing mutations arguing against a major role of mutations in the synaptotagmin XI gene in the pathogenesis of PD.

Introduction

The pathogenic causes of Parkinson's disease (PD) are widely unknown. Environmental and genetic causes are discussed as factors for the loss of dopaminergic neurons in the substantia nigra in affected individuals. In fact, there are an increasing number of genes in which mutations have been identified in autosomal recessive, autosomal dominant, and apparently sporadic patients. Altered gene products are thought to be involved in the ubiquitin-mediated protein degradation pathway (reviewed in Krüger et al., 2002). Two proteins, α-synuclein and parkin, play a key role in this pathway (Polymeropoulus et al., 1997; Kitada et al., 1998). Although the exact function of α-synuclein remains to be identified, there is solid evidence based on cell culture and transgenic studies that it is involved in synaptic transmission. The second known gene in PD, parkin, encodes a synaptic protein, which is linked to the vesicular transport system of neuronal cells (Kubo et al., 2001). It is localized in the Golgi complex of neuronal cells and on secretory vesicles together with synaptotagmin I and synaptophysin (Kubo et al., 2001). Moreover parkin has been implicated in proteasomal protein degradation via its ubiquitin-E3-ligase function (Shimura et al., 2000) which results in ubiquitination and subsequent degradation of the glycosylated form of α-synuclein (Shimura et al., 2001), synphilin 1 (Chung et al., 2001), CDCrel (Zhang et al., 2000), Pael-receptor (Imai et al., 2001), CHIP (Imai et al., 2002), HSP70 (Tsai et al., 2003), p38 subunit of aminocyl tRNA synthetase (Corti et al., 2003), cyclin E (Staropoli et al., 2003), and α/β-tubulin heterodimers (Ren et al., 2003), and binds CASK/LIN-2 (Fallon et al., 2001). Based on the genetic characterization of parkin-interacting proteins, we recently identified a novel mutation in the synphilin-1 gene responsible for PD in two apparently sporadic PD patients (Marx et al., 2003). This illustrates that physiological substrates of the E3-ligase parkin are good candidates for mutation screenings in sporadic and familial PD patients.

Recently, some of us identified synaptotagmin XI as another substrate of parkin (Huynh et al., 2003). Synaptotagmin XI belongs to the family of synaptotagmins. Members of this family are vesicle proteins, probably playing a role in the docking and fusion of synaptic vesicles with the plasma membrane and hence in transmitter release. Moreover synaptotagmins might be involved in endocytosis, since all representatives studied so far show high affinity for AP2, a clathrin adaptor protein complex (Ullrich et al., 1994; Zhang et al., 1994; Li et al., 1995). Synaptotagmins are found in abundance in nerve cells and some endocrine cells. Their amino acid sequence comprises a single transmembrane region and two cytoplasmic C2 domains (C2A and C2B) in the C-terminal region, followed by a conserved C-terminus (A-latrotoxin receptor interaction domain).

C2 domains are modules found in a number of cellular proteins and are important in signal transduction and membrane transport. The two synaptotagmin C2 domains have been shown to have different functions: C2A is responsible for Ca^{2+}-dependent binding to phospholipid (Davletov, 1993; Sutton, 1995), while C2B binds phospholipid and inositol polyphosphate in a calcium-independent manner (Fukuda, 1994).

Synaptotagmin XI is highly expressed in brain cells, but is also detectable in other tissues. Together with synaptotagmin IV, being most closely related to synaptotagmin XI, it constitutes a separate subgroup of synaptotagmins. Characteristic feature of this subgroup is an amino acid substitution (aspartate → serine) within the Ca^{2+}-binding site of the C2A domain that abolishes Ca^{2+}-dependent phospholipid binding. Members of this subclass of synaptotagmins are therefore expected to have Ca^{2+}-independent functions that await further elucidation (von Poser, 1997).

Synaptotagmin XI is located on chromosome 1q21.2 and thus not in any of the as yet known candidate regions or susceptibility loci for Parkinson associated genes. The gene has an approximate length of 5 kb, contains four exons and encodes a protein of 430 amino acids.

Being substrate for parkin and involved in the synaptic vesicle function synaptotagmin XI is an interesting candidate for PD. We therefore searched for mutations in the *synaptotagmin XI* gene in a large group of PD patients.

Material and methods

PD patients

A total of 393 sporadic and familial PD patients (mean age at disease onset 56.0 years (S.D. ± 11.7 years); males 55.1%, females 44.9%) were screened for mutations in the *synaptotagmin XI* gene. All patients were evaluated by a neurologist experienced in the diagnosis of PD based on the UK Parkinson's disease brain bank criteria. All subjects gave informed consent. For all PD patients, mutations in the alpha-synuclein and UCHL1-gene were excluded, and for juvenile PD patients also mutations in the Parkin- and DJ1-gene, respectively.

Primers and PCR conditions

PCR and sequencing primers (Table 1) were designed using Synaptotagmin XI genomic sequence information (Ensembl Gene ID #ENSG00000132718) and the primer 3 program (http://www-genome.wi.mit.edu/cgi-bin/primer/primer3_www.cgi) (Rozen and Skaletsky, 2000). Exon 2 had to be divided into three overlapping fragments being suitable for denaturing high performance liquid chromatography (dHPLC) analysis (designated exon 2_1, 2_2 and 2_3 in the following). For exon 1, 2-1, 2-2, 2-3, 3, and 4, 393, 167, 359, 350, 359, and 345 DNA samples were analyzed by dHPLC as PCR and dHPLC did not reveal an evaluatable result in all cases.

Table 1. PCR primers used for amplification of the 4 exons of the synaptotagmin XI gene and optimized temperature and elution conditions for mutation detection

	Name	Sequence (5′ → 3′)	PCR product	Annealing temperatures (T_A)	dHPLC column temperature (T_C)	dHPLC gradient [in %B]
Exon 1	Ex1_F Ex1_R	gtctcaccattgcaaaaacg ccttgcaaccattctgacc	131 bp	53°C	59°C	50–60
Exon 2_1	Ex2_8416_F Ex2_8803_R	aagtccacccgccctgac gtccattttgatgggtaattggtc	388 bp	55°C	60.0°C	60–70
Exon 2_2	Ex2_8728_F Ex2_9120_R	cctgctaagccgagacaaag gggatgccatagaaggtgaa	393 bp	56°C	58.0°C and 60°C	61–71 and 59–69
Exon 2_3	Ex2_9023_F Ex2_9320_R	accatccttcctgacaaacg acctaccattgctccccaac	298 bp	58°C	62.0°C	56–66
Exon 3	Ex3_F Ex3_R	tctgccaggattctcacc gatggttcagacagtaagaaatagg	203 bp	53°C	59.5°C	54–64
Exon 4	Ex4_F Ex4_R	tgtcataggtctgtctccctttt ccccggggattagaggag	387 bp	55°C	59.0°C	60–70

After amplification the best possible separation of homo- and heteroduplexes is achieved using the temperature and gradient parameters presented in this table. The acetonitril gradient (dHPLC gradient) is commonly indicated as the proportion of buffer B (25% (v/v) acetonitril) in the mobile phase. Exon 2_2 had to be analysed under two different temperature and elution conditions since two melting domains in the PCR product had arisen from the analysis of the melting profile (0–100 bp and 100–400 bp)

After optimizing PCR conditions by gradient PCR with a PTC-0200 DNA Engine (MJ Research, Inc.; Waltham, Mass.), PCR was performed on patient samples in a thermocycler (PTC-0200 DNA Engine, Perkin-Elmer 9600) and a final volume of 25 μl using 5 μl DNA, 1 μl of each primer (10 pmol/μl), 2.5 μl 10× PCR buffer (Genecraft), 0.5 μl 10 mM dNTPs and 1.5 U Taq polymerase (Genecraft) with the annealing temperatures given in Table 1. The following PCR conditions were applied: 2,5′ 94°C; [30″ 94°C; 30″ annealing temperature; 30″ 72°C] for 35 cycles; 5′ 72°C; 4°C ∝. Five μl of each PCR product were electrophoresed on 1% agarose gels to determine the presence and size of the amplified DNA. Only samples with positive PCR signals were pooled pair wise in the subsequent dHPLC analysis.

dHPLC

dHPLC (WAVE® DNA Fragment Analysis system [Transgenomic, Inc., San Jose, CA]) was used to screen for mutations in the *synaptotagmin XI* gene. Mutation detection relies on the different melting points of hetero- and homoduplexes. PCR products were pooled pairwise (10 μl each), subsequently denatured (94°C for 2 min), and slowly renatured (slow cooling to 10°C within 20 min) to allow the formation of heteroduplexes. Separation of homo- and heteroduplexes takes place in the cartridge of the wave system. DNA fragments bind to the stationary phase, a hydrophobic poly(styrene-divinylbenzene) bead matrix, via triethylammonium acetate (TEAA) as "bridging" molecule, single stranded fragments binding more tightly than double stranded fragments. The mobile phase consists of a buffer system composed of buffer A and B. Buffer A contains 0.1 M TEAA and 0.025% (v/v) acetonitrile (ACN) and buffer B contains 0.1 M TEAA and 25% (v/v) ACN. The hydrophobic interaction between stationary phase and TEAA is reduced with the increase of acetonitrile in the mobile phase. Under partially denaturing conditions (increased temperature) heteroduplexes are marginally more denatured than homoduplexes and elute therefore slightly before the homoduplexes. This difference is detected by the system's UV detector and can be seen in chromatograms, ideally as four distinct peaks (two homo- and two heteroduplexes). Complete resolution of the two homo- and heteroduplexes however is not obligatory. For more detailed information regarding the methodology please see (Huber et al., 1993; Kuklin et al., 1997–98).

The melting profiles of the PCR fragments under investigation were used to predict the temperature suitable for mutation scanning of the fragment. The ACN gradient (ratio of buffer A and B) was adjusted so that the retention times of homoduplexes lay between 1.6 and 1.9 minutes. Samples yielding conspicuous chromatograms were examined separately using the WAVE system and the samples to be considered were sequenced subsequently.

Gradient and temperature conditions suitable for mutation detection are presented in Table 1. These conditions have been confirmed for three different CEPH DNA samples.

Sequencing

After purifying PCR products using a PCR purification kit (Qiagen), both strands were sequenced using CEQ 2000 Dye Terminator Cycle Sequencing (Beckman Coulter). For each cycle sequencing reaction 200 ng of the amplicon and 12 pmol of primer (Table 1) were added to 3 μl of Quick Start Master Mix (Beckman Coulter) in a volume of 10 μl. The cycling profile was 96°C for 20 s, 50°C for 20 s and 60°C for 4 min for 30 cycles following a 90-s pre-denaturation at 94°C.

Cycle sequencing reactions were stopped with 5 μl of a mix containing one portion glycogen and two portions of each 100 mM EDTA (pH 8.0) and 3 M NaAc (pH 5.2) and precipitated with 60 μl 100% ethanol. After a 30-min spin (4°C, 16100 rpm) the pellet was washed twice with 70% ethanol, vacuum dried for 12.5 min and resuspended in 40 μl CEQ sample loading solution (Beckman Coulter). Sequences were obtained using a capillary sequencer (CEQ 8000, Beckman Coulter). Conditions included a capillary temperature of 50°C, 120″ denaturation at 90°C, 15″ injection at 2.0 kV and 85-min separation at 4.2 kV. Sequencing data were analysed with the CEQ 8000 analysis software.

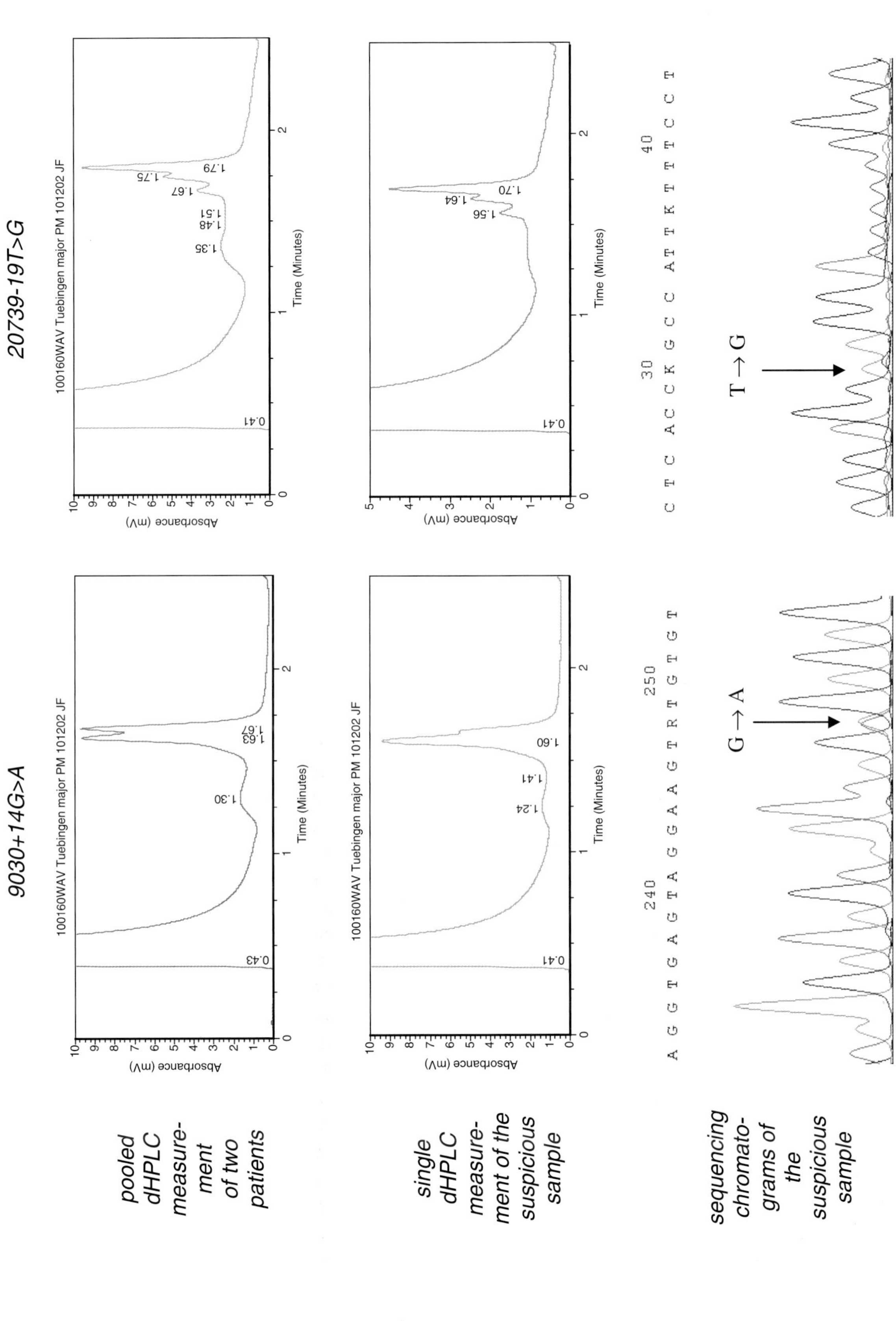

Fig. 1. dHPLC profiles and sequencing results of DNA of one patient revealing the two intronic mutations 9030 + 14G > A and 20739 − 19T > G

Results

dHPLC was performed to search for mutations in the *synaptotagmin XI* gene in up to 393 DNA samples of PD patients. Peak shapes typical for heteroduplexes appeared twice, namely in one PCR product of exon 2_3 and exon 3, respectively. No other obviously aberrant heteroduplex pattern was observed. Interestingly, both conspicuous patterns could be assigned to the same patient. Sequencing revealed two intronic mutations adjacent to exon 2_3 (9030 + 14G > A) and exon 3 (20739 – 19T > G), respectively (Fig. 1).

Other conspicuous peak shapes, such as "pre-peaks" and "shoulders" did not prove to be new sequence variants. However, sequencing of six representatives for exon 2_2 due to pre-peaks revealed an already described single nucleotide polymorphism (SNP) at genomic position 8715 (8715T > C) or cDNA position 546 (546T > C), respectively. This SNP does not change the amino acid composition and was observed in all six representatives independently of the pattern of pre-peaks.

Discussion

This is the first study to search for mutations in the *synaptotagmin XI* gene in PD. Synaptotagmin XI has been recently shown to interact with and serve as a substrate of Parkin, an E3 ubiquitin ligase (Shimura et al., 2000) which is commonly mutated in early onset autosomal recessive PD (Kitada et al., 1998). Therefore, and as the vesicle transport or release seems to be altered in PD, synaptotagmin XI was considered an excellent candidate for PD. Recently, for another synaptic vesicle associated protein that is ubiquitinated by parkin, synphilin-1, a novel mutation responsible for an apparently sporadic form of PD was identified (Marx et al., 2003). Synaptotagmin XI is highly expressed in brain (von Poser et al., 1997) and has been shown to be a component of the Lewy bodies (Huynh et al., 2003), a pathological hallmark of PD. Disturbed synaptotagmin XI function or an altered conformation could therefore be causative for the disease process for instance via increased release of dopamine into the cytoplasm resulting in increased oxidative damage (reviewed in Betarbet et al., 2002). This process might be coupled by the regulation of Ca^{2+} influx by syntaxin, SNAP-25, and synaptotagmin (reviewed in Atlas, 2001). Disturbed synaptic function seems to be important in the pathogenesis of PD, since several proteins known to be mutated in familial forms are closely related to synaptic integrity and function, i.e. α-synuclein, parkin and synphilin-1. Loss of α-synuclein function, for instance, leads to impaired response to prolonged, low-frequency stimulation, indicating disturbed vesicle recruitment from reserve pools (Cabin et al., 2002). In this context known synaptotagmin-XI-mediated down-regulation of presynaptic neurotransmission might be involved in dopaminergic dysfunction in PD (Wang et al., 2001).

However, in our study including a large sample of up to 393 PD patients, we did not identify any disease causing mutations. In contrast, synaptotagmin XI turned out to be highly conserved since no amino acid variations were found. This might reflect that peptide sequence variations of synaptotagmin XI dramatically alter its function and may not be compatible with physiological embryonic development. However, further investigations of the 5′UTR and promoter regions including functional analyses of detected sequence variations should be performed in further studies. Moreover, the generation of knock out animals is essential to prove this hypothesis.

References

Atlas D (2001) Functional and physical coupling of voltage-sensitive calcium channels with exocytotic proteins: ramifications for the secretion mechanism. J Neurochem 77: 972–985

Betarbet R, Sherer TB, Di Monte DA, Greenamyre JT (2002) Mechanistic approaches to Parkinson's disease pathogenesis. Brain Pathol 12: 499–510

Cabin DE, Shimazu K, Murphy D, Cole NB, Gottschalk W, McIlwain KL, Orrison B, Chen A, Ellis CE, Paylor R, Lu B, Nussbaum RL (2002) Synaptic vesicle depletion correlates with attenuated synaptic responses to prolonged repetitive stimulation in mice lacking alpha-synuclein. J Neurosci 22: 8797–8807

Chung KK, Zhang Y, Lim KL, Tanaka Y, Huang H, Gao J, Ross CA, Dawson VL, Dawson TM (2001) Parkin ubiquitinates the alpha-synuclein-interacting protein, synphilin-1: implications for Lewy-body formation in Parkinson disease. Nat Med 7: 1144–1150

Corti O, Hampe C, Koutnikova H, Darios F, Jacquier S, Prigent A, Robinson JC, Pradier L, Ruberg M, Mirande M, Hirsch E, Rooney T, Fournier A, Brice A (2003) The p38 subunit of the aminoacyl-tRNA synthetase complex is a Parkin substrate: linking protein biosynthesis and neurodegeneration. Hum Mol Genet 12: 1427–1437

Davletov BA, Sudhof TC (1993) A single C2 domain from synaptotagmin I is sufficient for high affinity Ca^{2+}/phospholipid binding. J Biol Chem 268(35): 26386–26390

Fallon L, Moreau F, Croft BG, Labib N, Gu WJ, Fon EA (2001) Parkin and CAK/LIN-2 associate via a PDZ-mediated interaction and are co-localized in lipid rafts and postsynaptic densities in brain. J Biol Chem 277: 486–491

Fukuda M, Aruga J, Niinobe M, Aimoto S, Mikoshiba K (1994) Inositol-1,3,4,5-tetrakisphosphate binding to C2B domain of IP4BP/synaptotagmin II. J Biol Chem 269(46): 29206–29211

Huber CG, Oefner PJ, Preuss E, Bonn GK (1993) High-resolution liquid chromatography of DNA fragments on non-porous poly(styrene-divinylbenzene) particles. Nucl Acids Res 21: 1061–1066

Huynh DP, Scoles DR, Nguyen D, Pulst SM (2003) The autosomal recessive juvenile Parkinson disease gene product, parkin, interacts with and ubiquitinates synaptotagmin XI. Hum Mol Genet 12: 1–11

Imai Y, Soda M, Inoue H, Hattori N, Mizuno Y, Takahashi R (2001) An unfolded putative transmembrane polypeptide, which can lead to endoplasmic reticulum stress, is a substrate of Parkin. Cell 105: 891–902

Imai Y, Soda M, Hatakeyama S, Akagi T, Hashikawa T, Nakayama KI, Takahashi R (2002) CHIP is associated with Parkin, a gene responsible for familial Parkinson's disease, and enhances its ubiquitin ligase activity. Mol Cell 10: 55–67

Kitada T, Asakawa S, Hattori N, Matsumine H, Yamamura Y, Minoshima S, Yokochi M, Mizuno Y, Shimizu N (1998) Mutations in the parkin gene cause autosomal recessive juvenile parkinsonism. Nature 392: 605–608

Kubo SI, Kitami T, Noda S, Shimura H, Uchiyama Y, Asakawa S, Minoshima S, Shimizu N, Mizuno Y, Hattori N (2001) Parkin is associated with cellular vesicles. J Neurochem 78: 42–54

Kruger R, Eberhardt O, Riess O, Schulz JB (2002) Parkinson's disease: one biochemical pathway to fit all genes? Trends Mol Med 8: 236–240

Kuklin A, Munson K, Gjerde D, Haefele R, Taylor P (1997–98) Detection of single-nucleotide polymorphisms with the WAVE DNA fragment analysis system. Genet Test 1: 201–206

Li C, Ullrich B, Zhang JZ, Anderson RG, Brose N, Sudhof TC (1995) Ca(2+)-dependent and -independent activities of neural and non-neural synaptotagmins. Nature 375: 594–599

Marx FP, Holzmann C, Strauss KM, Li L, Eberhardt O, Gerhardt E, Cookson MR, Hernandez D, Farrer MJ, Kachergus J, Engelender S, Ross CA, Berger K, Schöls L, Schulz JB, Riess O, Kruger R (2003) Identification and functional characterization of a novel R621C mutation in the synphilin-1 gene in Parkinson's disease. Hum Mol Genet 12: 1223–1231

Ren Y, Zhao J, Feng J (2003) Parkin binds to α/β-tubulin and increases their ubiquitination and degradation. J Neurosci 23: 3316–3324

Rozen S, Skaletsky HJ (2000) Primer 3 on the WWW for general users and for biologist programmers. In: Krawetz S, Misener S (eds) Bioinformatics methods and protocols: methods in molecular biology. Humana Press, Totowa, pp 365–386

Schlossmacher MG, Frosch MP, Gai WP, Medina M, Sharma N, Forno L, Ochiishi T, Shimura H, Sharon R, Hattori N (2000) Parkin localizes to the Lewy bodies of Parkinson disease and dementia with Lewy bodies. Am J Pathol 160: 1655–1667

Shimura H, Hattori N, Kubo S, Mizuno Y, Asakawa S, Minoshima S, Shimizu N, Iwai K, Chiba T, Tanaka K (2000) Familial Parkinson disease gene product, parkin, is a ubiquitin-protein ligase. Nat Genet 25: 302–305

Shimura H, Schlossmacher MG, Hattori N, Frosch MP, Trockenbacher A, Schneider R, Mizuno Y, Kosik KS, Slekoe DJ (2001) Ubiquitination of a new form of alpha-synuclein by parkin from human brain: implications for Parkinson's disease. Science 293: 263–269

Staropoli JF, McDermott C, Martinat C, Schulman B, Demireva E, Abeliovich A (2003) Parkin is a component of an SCF-like ubiquitin ligase complex and protects postmitotic neurons from kainite excitotoxicity. Neuron 37: 735–749

Sutton RB, Davletov BA, Berghuis AM, Sudhof TC, Sprang SR (1995) Structure of the first C2 domain of synaptotagmin I: a novel Ca^{2+}/phospholipid-binding fold. Cell 80(6): 929–938

Tsai YC, Fishman PS, Thakor NV, Oyler GA (2003) Parkin facilitates the elimination of expanded polyglutamine proteins and leads to preservation of proteasome function. J Biol Chem 278: 22044–22055

Ullrich B, Li C, Zhang JZ, McMahon H, Anderson RG, Geppert M, Sudhof TC (1994) Functional properties of multiple synaptotagmins in brain. Neuron 13: 1281–1291

Von Poser C, Ichtchenko K, Shao X, Rizo J, Südhof TC (1997) The evolutionary pressure to inactivate. A subclass of synaptotagmins with an amino acid substitution that abolishes Ca^{2+} binding. J Biol Chem 272: 14314–14319

Wang CT, Grishanin R, Earles CA, Chang PY, Martin TF, Chapman ER, Jackson MB (2001) Synpatotagmin modulation of fusion pore kinetics in regulated exocytosis of dense-core vesicles. Science 294: 1111–1115

Zhang JZ, Davletov BA, Sudhof TC, Anderson RG (1994) Synaptotagmin I is a high affinity receptor for clathrin AP-2: implications for membrane recycling. Cell 78: 751–760

Zhang Y, Gao J, Chung KK, Huang H, Dawson VL, Dawson TM (2000) Parkin functions as an E2-dependent ubiquitin-protein ligase and promotes the degradation of the synaptic vesicle-associated protein, CDCrel-1. Proc Natl Acad Sci USA 97: 13354–13359

Authors' address: O. Riess, MD, Department of Medical Genetics, University of Tübingen, Calwerstrasse 7, D-72070 Tübingen, Germany, e-mail: olaf.riess@med.uni-tuebingen.de

J Neural Transm (2004) [Suppl] 68: 29–38

Depletion and restoration of endogenous monoamines affects β-CIT binding to serotonin but not dopamine transporters in non-human primates

A. Heinz[1], **D. W. Jones**[2], **K. Zajicek**[3], **J. G. Gorey**[2], **G. Juckel**[1], **J. D. Higley**[3], and **D. R. Weinberger**[2]

[1] Department of Psychiatry, Charité University Medicine Berlin, Campus Charité Mitte, Berlin, Germany
[2] Clinical Brain Disorders Branch, NIMH, NIH, and
[3] Laboratory of Clinical Studies, DICBR, NIAAA, Bethesda, MD, USA

Summary. The radioligand [^{123}I]β-CIT binds to dopamine transporters in striatum and to serotonin transporters in brainstem. Endogenous dopamine or serotonin may compete with radioligand binding at monoamine transporters.

We used α-methyl-p-tyrosine (AMPT) to block dopamine production and measured [^{123}I]β-CIT binding before and after endogenous dopamine was restored by IV administration of the dopamine precursor L-dihydroxyphenylalanine (L-DOPA) in rhesus monkeys. P-chlorophenylalanine (pCPA) was used to inhibit serotonin production, and [^{123}I]β-CIT binding was assessed before and after IV administration of the serotonin precursor 5-hydroxy-L-tryptophan (L-5-HTP) restored endogenous serotonin. Pretreatment with benserazide blocked peripheral decarboxylization in both paradigms.

Serotonin restoration measurably displaced [^{123}I]β-CIT binding to brainstem serotonin transporters but not to striatal dopamine transporters. Restoration of dopamine apparently did not affect [^{123}I]β-CIT binding to striatal dopamine transporters. However, dopamine restoration reduced radioligand binding to brainstem serotonin transporters, most likely due to dopamine release from serotonin neurons following L-DOPA administration.

The higher striatal density of dopamine transporters relative to dopamine concentrations may explain why [^{123}I]β-CIT displacement by endogenous dopamine was not observed. This study indicates that [^{123}I]β-CIT binding in brainstem (raphe area) is affected by endogenous serotonin release *in vivo* and that L-DOPA treatment may cause serotonin neurons in the brainstem to corelease dopamine.

Introduction

The radioligand [^{123}I]β-CIT is widely used with SPECT to measure the availability of central dopamine and serotonin transporters in neurological and psychiatric disorders such as Parkinson's disease, alcoholism and major depression (Rinne et al., 1995; Heinz et al., 1998; Malison et al., 1998). Monoamine transporters regulate the reuptake of released neurotransmitters, and neurotransmitter concentrations measured *in vivo* with microdialysis were negatively correlated with the availability of monoamine transporters measured with [^{123}I]β-CIT and SPECT (Heinz et al., 1999). However, the observed negative correlation between neurotransmitter concentrations and [^{123}I]β-CIT binding could also be caused by competition between the endogenous neurotransmitter and the radioligand for transporter binding sites. Fisher et al. (1995) suggested that endogenous monoamines compete with SPECT and PET radioligands for receptor or transporter binding. Dopamine depletion caused by enzymatic blockade of endogenous dopamine production was indeed associated with an increase in binding of the radioligand IBZM to dopamine D2 receptors (Abi-Dargham et al., 2000). This finding supports the hypothesis that radioligand binding reflects the "availability" of D2 receptors that are currently not occupied by endogenous dopamine rather than the absolute density of D2 receptors. However, so far it has not been examined whether endogenous monoamines also compete with [^{123}I]β-CIT binding to transporter sites. This question is not only of scientific interest but may have diagnostic implications: if dopamine competes with [^{123}I]β-CIT binding to dopamine transporters, then low dopamine concentrations in Parkinson's disease may be associated with higher radioligand binding, resulting in an underestimation of dopaminergic dysfunction. Therefore, we blocked either dopamine or serotonin production in rhesus monkeys and examined whether restoration of the endogenous neurotransmitter would displace the radioligand [^{123}I]β-CIT from its dopamine or serotonin transporter binding sites. SPECT imaging with [^{123}I]β-CIT has been very well characterized in previous studies of human and non-human primates. These studies have established its steady-state stability at extended times (16–24 hr post injection) due to sustained plasma concentrations, and its off-rate (k*off*) in response to competing pharmaceutical challenges (Laruelle et al., 1993, 1994; Pirker et al., 1995, 2000). Based on this solid foundation of prior validation, we decided that this technique was ideal for testing our hypotheses in this pilot study.

Methods

All protocols used in this study were reviewed and approved by animal care committees of the intramural research programs of the NIH. We examined two rhesus monkeys (*macaca mulatta*, age five years) with 11 kg body weight. Concentrations of the serotonin metabolite, 5-hydroxyindoleacteic acid (HIAA) and the dopamine metabolite homovanillic acid (HVA) in the cerebrospinal fluid (CSF) were acquired in the week before the SPECT scan via cisternal puncture within 25 minutes of ketamine (10 mg/kg i.m.) administration and quantified with liquid chromatography and electrochemical-detection as described in detail elsewhere (Heinz et al., 2003). As previously reported (Heinz et al., 1998), CSF 5-HIAA concentrations measured at age two and three in eleven rhesus monkeys were significantly correlated with 5-HIAA levels collected at age five ($R = 0.65$ and $R = 0.77$, $p < 0.05$, resp.), indicating a trait-like character of serotonergic neurotransmission in these primates.

Dopamine transporters in the striatum and serotonin transporters in the raphe area were assessed with [^{123}I]β-CIT and SPECT as described previously (Heinz et al., 1999, 2003). [^{123}I]β-CIT binds with high affinity to striatal dopamine transporters and to serotonin transporters

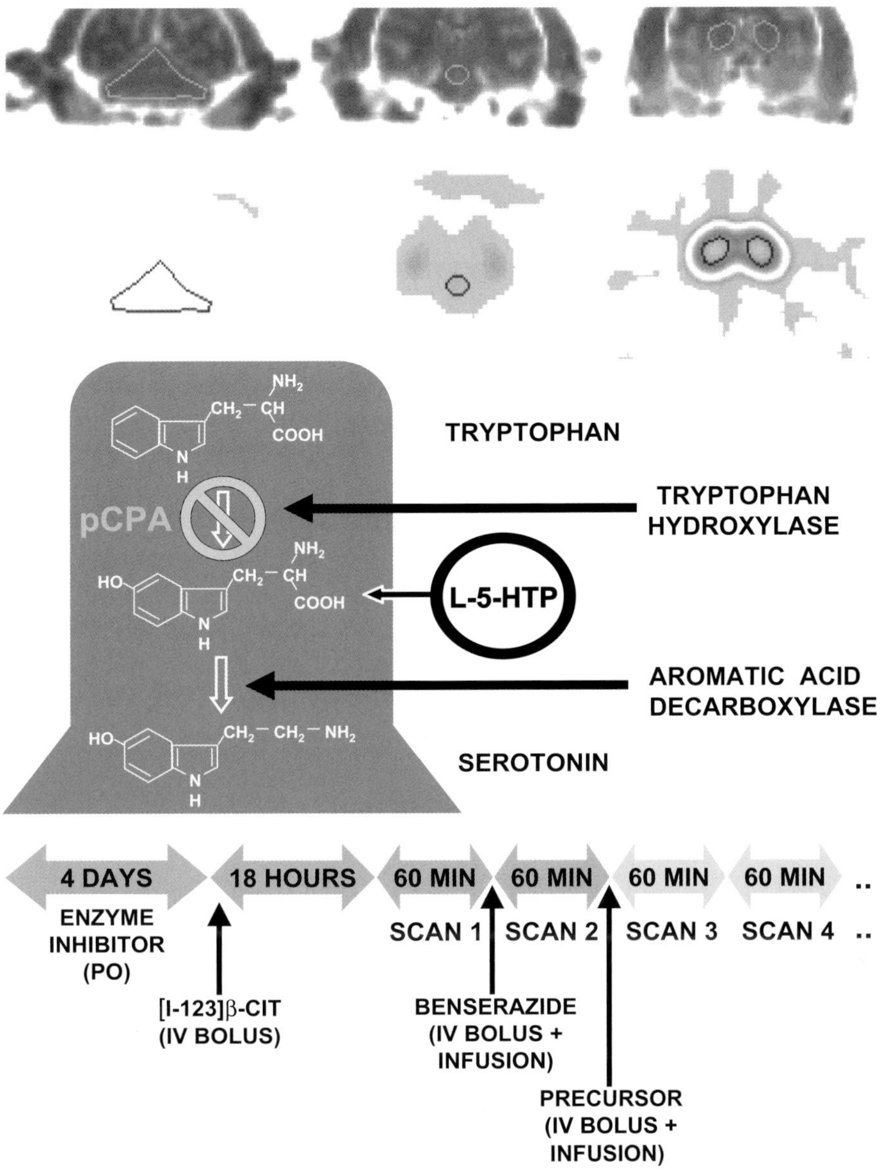

Fig. 1. Study design. *Top*: Placement of regions of interest (ROIs) for the cerebellum, brainstem and striatum on coregistered MRIs and SPECT scans. *Middle*: Biochemical interventions for serotonin depletion and restoration: Serotonin production via tryptophan hydroxylase is blocked by p-chlorophenylalanine (pCPA); afterwards, serotonin is restored by administration of 5-hydroxy-L-tryptophan (L-5-HTP), which is converted to serotonin by the aromatic acid decarboxylase. *Bottom*: Study design for SPECT imaging: Serotonin production is blocked by the enzyme inhibitor pCPA ($\approx$300 mg/kg/day) for 4 days. The radioligand β-CIT is applied and a baseline 60-minute scan is acquired 18 hours later, when pseudoequilibrium is established at dopamine and serotonin binding sites. Benserazide is administered to reduce peripheral metabolization of the serotonin precursor L-5-HTP (100 mg/kg), which is administered by bolus and infusion (1:1). Effects of intrasynaptic serotonin restoration are measured with several consecutive 60-minute scans

in the brainstem (Seibyl et al., 1994). [^{123}I]β-CIT was prepared (Baldwin et al., 1993) and administered intravenously [dose range: 15–21 mCi]. A series of 60-minute SPECT scans [CERASPECT; Digital Scintigraphics, Waltham, MA; full width at half-maximum 7.5 mm;

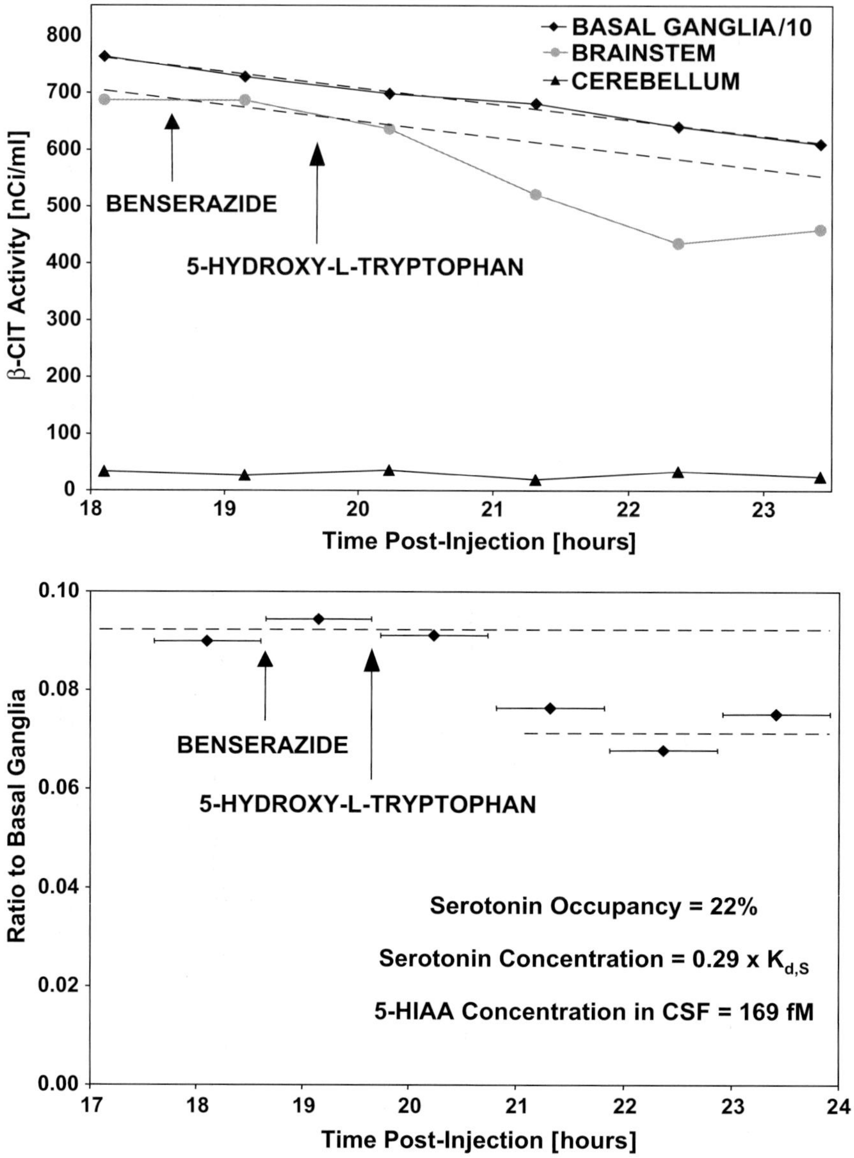

Fig. 2. Serotonin restoration and β-CIT binding in monkey #1. *Top*: Restoration of serotonin was associated with a 22% decrease in β-CIT binding in the brainstem, hypothetically reflecting competition of endogenous serotonin with the radioligand β-CIT for binding at brainstem serotonin transporters. In the striatum, β-CIT mainly binds to dopamine transporters (Laruelle et al., 1993), and no significant effect of serotonin restoration on striatal radioligand binding was observed. Please note that β-CIT binding in the basal ganglia is much higher compared with the brainstem, so that measured radioactivity (nCi/ml) in the basal ganglia was divided by 10 to fit into the same scale as the activity measured in the brainstem and cerebellum. *Bottom*: After serotonin depletion, the ratio of β-CIT binding in the brainstem ROI to striatal β-CIT uptake was measured before and after serotonin restoration. The ratio of brainstem to striatal β-CIT binding before serotonin restoration (average of the first two 60-minute scans) is supposed to reflect the density (B*max*) of brainstem serotonin transporters. After serotonin restoration, this ratio (average of the last two 60-minute scans) reflects the availability of serotonin transporters (B*avail*) that are currently not occupied by endogenous serotonin and available for radioligand binding

128×128×64 1.67 mm contiguous, isotropic voxels per scan] were acquired 18 hours after injection, when equilibrium is established at brainstem dopamine and serotonin transporters (Laruelle et al., 1993; Pirker et al., 2000). Individual ROIs were drawn based on a standard atlas of rhesus macaque brain and comprised the basal ganglia, brainstem, and cerebellum (Heinz et al.,

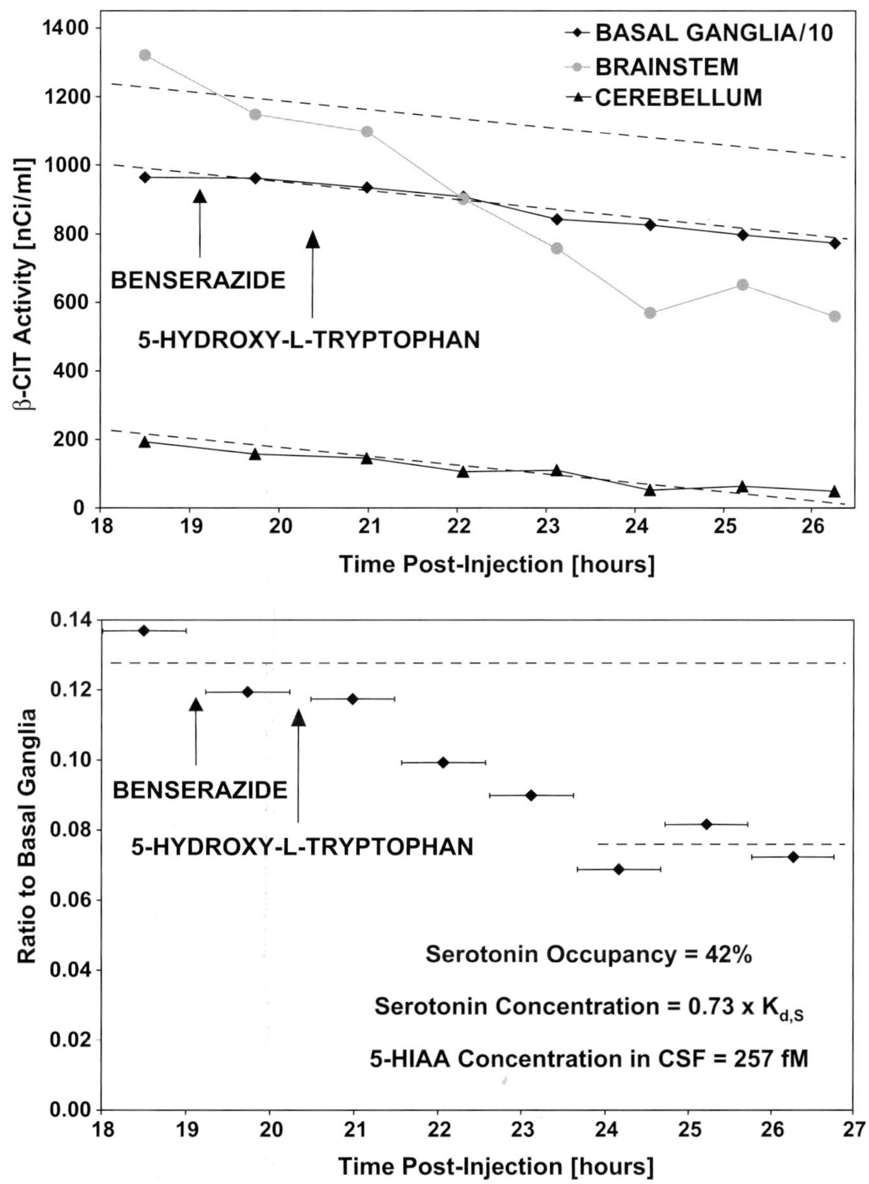

Fig. 3. Serotonin restoration and β-CIT binding in monkey #2. *Top*: Restoration of serotonin was associated with a 42% decrease in β-CIT binding in the brainstem. Again, no significant effect of serotonin restoration on striatal radioligand binding was observed. Please note that measured radioactivity (nCi/ml) in the basal ganglia was divided by 10 to fit into the same scale as the activity measured in the brainstem and cerebellum. *Bottom*: After serotonin depletion, the ratio of β-CIT binding in the brainstem ROI to striatal β-CIT uptake was measured before (average of the first two 60-minute scans; reflects B*max*) and after serotonin restoration (average of the last three 60-minute scans, reflects B*avail*)

1999, 2003; see Fig. 1). The highest density of dopamine transporters is found in the striatum (Laruelle et al., 1993) and of serotonin transporters in the brainstem (Jagust et al., 1996), which contains the raphe system. ROI placement was confirmed on coregistered MRI scans (Heinz et al., 1999). Activity concentration (nCi/ml) was measured for the volume encompassed by each set of ROIs.

In the first study, the two rhesus monkeys underwent oral administration for 4 consecutive days of approximately 300 mg/kg/day α-methyl-p-tyrosine (AMPT) to deplete dopamine by specifically blocking tyrosine hydroxylase (Kraemer and McKinney, 1979). The radioligand $[^{123}I]$β-CIT was injected and 18 hours later, a first baseline scan was acquired. Then the monkeys received 10 mg/kg benserazide i.v. (bolus:infusion = 1:1) to reduce peripheral decarboxylation of dopamine precursors and a second 60-minute baseline scan was acquired. Afterwards, endogenous dopamine was restored by infusion of the dopamine precursor L-dihydroxyphenylalanine (L-DOPA, 100 mg/kg; bolus:infusion = 1:1), and several 60-minute scans were acquired to examine potential displacement of the radioligand by restored endogenous dopamine. In a second study, the same protocol was followed, except that serotonin depletion was achieved with approximately 300 mg/kg/day p-chlorophenylalanine (pCPA) given orally for 4 consecutive days to deplete serotonin by specifically blocking tryptophan hydroxlase (Gradwell et al., 1975; Kraemer and McKinney, 1979); endogenous serotonin was restored with the serotonin precursor 5-hydroxy-L-tryptophan (L-5-HTP, 100 mg/kg; bolus:infusion = 1:1) to examine potential displacement of the radioligand by endogenous serotonin.

Striatal and brainstem binding of $[^{123}I]$β-CIT is often analyzed using the ratio of specific to nonspecific binding, measured when pseudoequilibrium is established at binding sites (Laruelle et al., 1994). However, more than 18 hours after radioligand application, nonspecific binding in the cerebellum is very low and difficult to measure accurately (see Fig. 1), resulting in increased variance when ROI activity is divided by cerebellar binding. Binding to striatal dopamine transporters, on the other hand, was stable at these late time points and not affected by monoamine depletion (Figs. 2 and 3, N.B. in these figures striatal data has been divided by ten for clarity), so we assessed the ratio of brainstem to striatal binding. Binding ratios measured after serotonin depletion (before restoration) are assumed to approximate B*max*, the density of serotonin transporters in the brainstem; however, this may somewhat underestimate B*max* if serotonin depletion is not complete (Gradwell et al., 1975). Binding ratios measured after serotonin restoration hypothetically reflect B*avail*, i.e. serotonin transporters not occupied by endogenous neurotransmitter and available for radioligand binding. Percent occupation of serotonin transporters by endogenous serotonin is determined as $\{1 - B_{avail}/B_{max}\}$ (eqn. 1); please note that this ratio is independent of striatal binding, which cancels out in the numerator and denominator). If the dissociation constant $K_{D,5\text{-}HT}$ of serotonin at brainstem serotonin transporters is known, then synaptic serotonin concentration is determined as $\{(B_{max}/B_{avail} - 1) \times K_{D,5\text{-}HT}\}$ (eqn. 2).

Results

In the monkey with lower pre-depletion 5-HIAA concentration in the CSF (169 fM), restoration of serotonin production was associated with a 22% decrease (eqn. 1) in β-CIT binding in the raphe area (Fig. 2, N.B. in this figure striatal data has been divided by ten for clarity). According to eqn. 2, estimated synaptic serotonin concentration was $0.29 \times K_{D,5\text{-}HT}$ in this monkey. In the monkey with higher pre-deletion 5-HIAA concentrations in the CSF (257 fM), restoration of serotonin production was associated with a 42% decrease (eqn. 1) in β-CIT binding in the raphe area (Fig. 3, N.B. in this figure striatal data has been divided by ten for clarity) and estimated synaptic serotonin concentration of $0.73 \times K_{D,5\text{-}HT}$. Neither β-CIT binding to dopamine transporters in striatum nor nonspecific β-CIT uptake in cerebellum was affected by serotonin restoration. Striatal or cerebellar binding of the radioligand β-CIT was not affected by dopamine restoration; however, a 30% decrease in β-CIT binding in the brain-

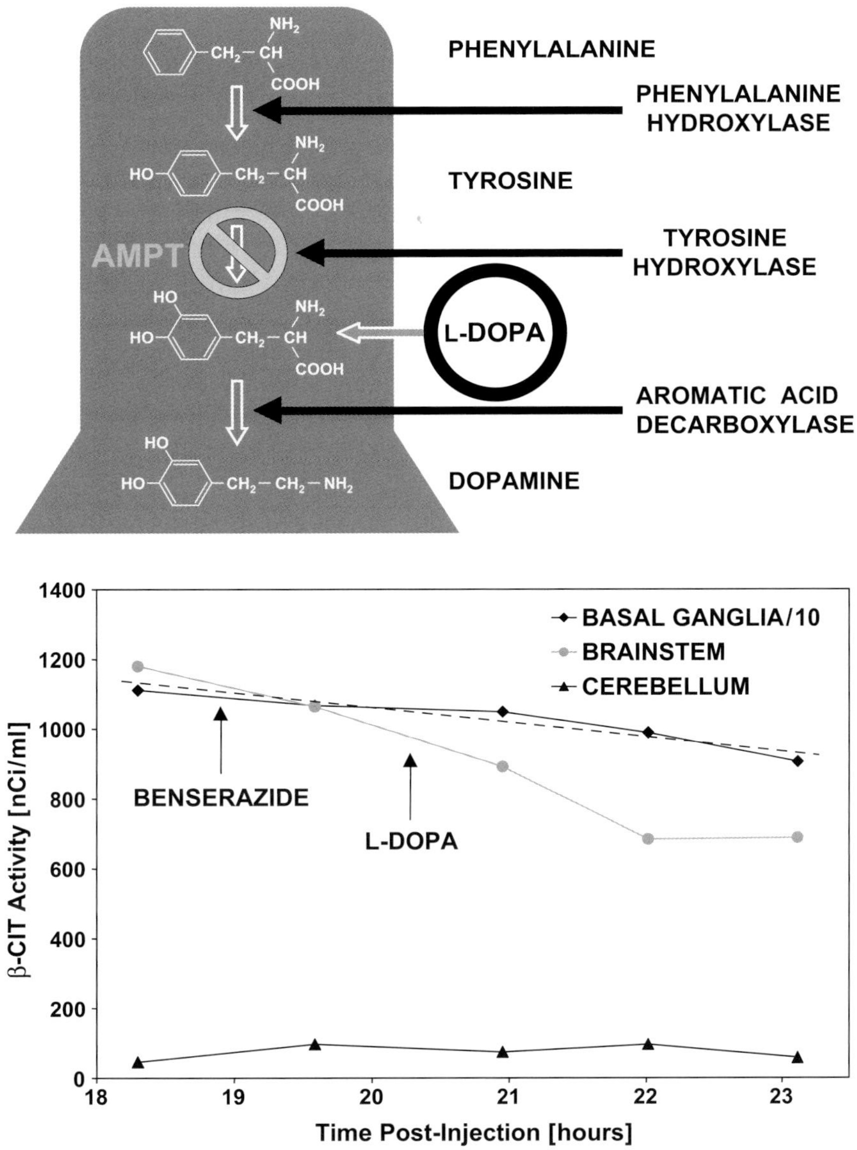

Fig. 4. Dopamine restoration and β-CIT binding in monkey #2. *Top*: Biochemical interventions for dopamine depletion and restoration: Dopamine production via tyrosine hydoxylase is blocked by α-methyl-p-tyrosine (AMPT); afterwards, dopamine is restored by application of L-dihydroxyphenylalanine (L-DOPA), which is converted to dopamine by the aromatic acid decarboxylase. *Bottom*: Striatal binding of the radioligand β-CIT was not affected by dopamine restoration; however, a 30% decrease in β-CIT binding in the brainstem was observed. Please note that β-CIT binding to the basal ganglia is much higher compared with brainstem binding, so that measured radioactivity (nCi/ml) in the basal ganglia was divided by 10 to fit into the same scale as measured activity in the brainstem and cerebellum

stem was observed in one monkey (Fig. 4, N.B. in this figure striatal data has been divided by ten for clarity). Due to technical problems, we were not able to complete the dopamine depletion and restoration phase of the study in the second primate.

Discussion

The preliminary data from this pilot study showed no evidence that $[^{123}I]\beta$-CIT binding in the basal ganglia is affected by endogenous dopamine release. It seems most likely that this failure to observe displacement of β-CIT binding by endogenous dopamine in the striatum is a consequence of a high density of dopamine transporters relative to tonic, average dopamine concentration in this area. There appears always to be a large reserve of unoccupied, available transporters that maintain low concentrations of extra-neuronal dopamine. As a result, B_{avail} is on average nearly equal to B_{max}, and occupancy by endogenous dopamine is difficult to measure *in vivo*. However, $[^{123}I]\beta$-CIT binding in the brainstem (raphe nuclei) was affected measurably by endogenous serotonin release *in vivo* and can be quantitatively assessed. Previous studies indicated that the applied dose of pCPA effectively blocked serotonin production (Kraemer and McKinney, 1979); if so, the ratio of transporter density (measured after serotonin depletion in the absence of competition with endogenous serotonin) to transporter availability (measured after endogenous serotonin was restored) can be used to assess synaptic serotonin concentrations. In fact, the ratio of estimated serotonin concentrations, $0.73/0.29 = 2.5$, measured non-invasively with *in vivo* SPECT agrees moderately well with the ratio of 5-HIAA in CSF, $257/169 = 1.5$, for the two monkeys in this pilot study. Similar procedures have been used to assess endogenous dopamine concentrations *in vivo* in schizophrenic patients and healthy volunteers by post-synaptic dopamine receptor availability rather than pre-synaptic transporters (Abi-Dargham et al., 2000).

At first, we were surprised to observe a reduction of $[^{123}I]\beta$-CIT binding in the brain stem as a result of endogenous dopamine restoration. In the brainstem, $[^{123}I]\beta$-CIT is displaced by selective ligands that bind to serotonin transporters, whereas in the striatum, $[^{123}I]\beta$-CIT is displaced by selective ligands that bind to dopamine transporters (Laruelle et al., 1993). In accordance with this observation, $[^{123}I]\beta$-CIT binding to brainstem serotonin transporters was inversely correlated with concentrations of the serotonin metabolite 5-hydroxyindoleacetic acid (5-HIAA) but not the dopamine metabolite homovanillic acid (HVA) in the cerebrospinal fluid of non-human primates and humans (Heinz et al., 1998, 2003). However, it is unlikely that either dopamine or serotonin is exclusively selective for its respective transporter, i.e. both these structurally related monoamines probably have considerable affinity for both types of transporters. Indeed, both dopamine uptake by norepinephrine transporters and norepinephrine uptake by dopamine transporters have been documented (Carboni et al., 1990; Giros et al., 1994; Moron et al., 2002; Torres et al., 2003). Similarly, serotonin uptake by dopamine transporters has been demonstrated (Pan et al., 2001; Suarez-Roca and Cubeddu, 2002; Zhou et al., 2002); we are unaware, however, of prior reports of dopamine interacting directly with serotonin transporters. L-DOPA, which was used to restore endogenous dopamine, crosses the blood brain barrier well and is taken up by all neurons, not just dopamine neurons. Thus, L-DOPA certainly enters serotonin neurons and will be decarboxylated into dopamine since the aromatic acid decarboxylase enzyme is common to both types of neurons. Extraneous dopamine produced by this process will be co-released with serotonin as a false transmitter, and apparently may compete effectively with serotonin at the serotonin transporter leading to elevated serotonin levels. Further, this extraneous dopamine may not be metabolized efficiently since it is typically not found in this area of the brain. Alternatively, displacement of $[^{123}I]\beta$-CIT in the brainstem may be caused by stimulated serotonin release rather than extraneous dopamine release. Activation of dopamine neurons in the ventral tegmentum area, for example, can stimulate serotonin release in the raphe nuclei: administration of the D_1/D_2 agonist apomorphine increased somatodendritic serotonin release in the dorsal raphe via stimulation of projection fibers to the raphe nuclei (Ferre and Artigas, 1993; Martin-Ruiz et al., 2001). In either case, L-DOPA treatment may have unwanted consequences as an indirect serotonin "agonist," and our study may be the first *in vivo* measurements of a mechanism underlying this phenomenon.

Altogether, this pilot study supports demonstrates the feasibility of using *in vivo* imaging techniques to examine endogenous neurotransmitters and their interactions with binding sites and one another. In the right circumstances, trace doses of radioligand can be used to assess quantitatively neurotransmitter concentrations non-invasively *in vivo*. Furthermore, our measurements suggest that L-DOPA administration may have unexpected serotonergic side-effects.

Acknowledgements

The authors thank Dr. K. A. Frey for a very useful discussion of the interpretation of our data, and the late Dr. M. Linnoila for his essential involvement in this project. Supported by the Deutsche Forschungsgemeinschaft (He 2597/7–1).

References

Abi-Dargham A, Rodenhiser J, Printz D, Zea-Ponce Y, Gil R, Kegeles LS, Weiss R, Cooper TB, Mann JJ, Van Heertum RL, Gorman JM, Laruelle M (2000) Increased baseline occupancy of D2 receptors by dopamine in schizophrenia. Proc Natl Acad Sci USA 97: 8104–8109

Baldwin RM, Zea-Ponce Y, Zoghbi SS, Laruelle M, Al-Tikritti MS, Sybirska EH, Malison RT, Neumeyer JL, Milius RA, Wang S, Stabin M, Smith EO, Charney DS, Hoffer PB, Innis RB (1993) Evaluation of the monoamine uptake site ligand $[^{123}I]$methyl-3β-(Iodophenyl)-tropane-2β-carboxylate ($[^{123}I]\beta$-CIT) in non-human primates: pharmacokinetics, biodistribution and SPECT brain imaging coregistered with MRI. Nucl Med Biol 5: 597–606

Carboni E, Tanda GL, Frau R, Dichiara G (1990) Blockade of the noradrenaline carrier increases extracellular dopamine concentrations in the prefrontal cortex – evidence that dopamine is taken up in vivo by noradrenergic terminals. J Neurochem 55: 1067–1070

Ferre S, Artigas F (1993) Dopamine D2 receptor-mediated regulation of serotonin extracellular concentration in the dorsal raphe nucleus of freely moving rats. J Neurochem 61: 772–775

Fisher RE, Morris ED, Alpert NM, Fischman AJ (1995) In vivo imaging of neuromodulatory synaptic transmission using PET: a review of relevant neurophysiology. Hum Brain Map 3: 24–34

Giros B, Wang YM, Suter S, Mcleskey SB, Pifl C, Caron MG (1994) Delineation of discrete domains for substrate, cocaine, and tricyclic antidepressant interactions using chimeric dopamine-norepinephrine transporters. J Biol Chem 269: 15985–15988

Gradwell PB, Everitt BJ, Herbert J (1975) 5-hydroxytryptamine in the central nervous system and sexual receptivity of female rhesus monkeys. Brain Res 88: 281–293

Heinz A, Higley JD, Jones DW, Gorey JG, Hommer D, Saunders R, Zajicek K, Suomi SS, Knable MB, Lesch KP, Weinberger DR, Linnoila M (1998) In vivo association between brainstem serotonin transporters and sensitivity to alcohol intoxication. Am J Psychiatry 155: 1023–1028

Heinz A, Saunders RC, Kolachana BS, Bertolino A, Jones DW, Gorey JG, Bachevalier J, Lee KS, Knable MB, Saunders RC, Weinberger DR (1999) Disinhibition of subcortical dopaminergic neurotransmission in rhesus monkeys with neonatal mesial temporal lesions. Synapse 39: 71–79

Heinz A, Jones DW, Gorey JG, Bennet A, Suomi SJ, Weinberger DR, Higley JD (2003) Serotonin transporter availability correlates with alcohol intake in non-human primates. Mol Psychiatry 8: 231–234

Jagust WJ, Eberling JL, Biegon A, Taylor SE, Van Brocklin HF, Jordan S, Hanrahan SM, Roberts JA, Brennan KM, Mathis CA (1996) Iodine-123-5-iodo-6-nitroquipazine: SPECT radiotracer to image the serotonin transporter. J Nucl Med 37: 1207–1214

Kraemer GW, McKinney WT (1979) Interactions of pharmacological agents which alter biogenic amine metabolism and depression. J Affect Disord 1: 33–54

Laruelle M, Baldwin RM, Malison RT, Zea-Ponce Y, Zoghbi SS, Al-Tikriti MS, Sybirska MH, Zimmermann RC, Wiesniewski G, Neumeyer JL, Milius RA, Wang S, Smith EO, Roth RH, Charney DS, Hoffer PB, Innis RB (1993) SPECT imaging of dopamine and serotonin transporters with [^{123}I]β-CIT: pharmacological characterization of brain uptake in non-human primates. Synapse 13: 295–309

Laruelle M, Wallace E, Seibyl JP, Baldwin RM, Zea-Ponce Y, Zoghbi SS, Neumeyer JL, Charney DS, Hoffer PB, Innis RB (1994) Graphical, kinetic, and equilibrium analyses of in vivo [^{123}I]β-CIT binding to dopamine transporters in healthy human subjects. J Cereb Blood Flow Metab 14: 982–994

Malison RT, Price LH, Berman R, van Dyck CH, Pelton GH, Carpenter L (1998) Reduced brain serotonin transporter availability in major depression as measured by [^{123}I]-2β-carboxy-3β-(4-iodophenyl)tropane and single photon emission computed tomography. Biol Psychiatry 44: 1090–1098

Martin-Ruiz R, Ugedo L, Honrubia MA, Mengod G, Artigas F (2001) Control of serotonergic neurons in rat brain by dopaminergic receptors outside the dorsal raphe nucleus. J Neurochem 77: 762–775

Moron JA, Brockington A, Wise RA, Rocha BA, Hope BT (2002) Dopamine uptake through the norepinephrine transporter in brain regions with low levels of the dopamine transporter: evidence from knock-out mouse lines. J Neurosci 22: 389–395

Pan Y, Gembom E, Peng W, Lesch K-P, Mossner R, Simantov R (2001) Plasticity in serotonin uptake in primary neuronal cultures of serotonin transporter knockout mice. Dev Brain Res 126: 125–129

Pirker W, Asenbaum S, Kasper S, Walter H, Angelberger P, Koch G, Pozzera A, Deecke L, Podreka I, Bruecke T (1995) β-CIT SPECT demonstrates blockade of 5HT-uptake sites by citalopram in the human brain in vivo. J Neural Transm 100: 247–256

Pirker W, Asenbaum S, Hauk M, Kandlhofer S, Tauscher J, Willeit M (2000) Imaging serotonin and dopamine transporters with ^{123}I-β-CIT SPECT: binding kinetics and effects of normal aging. J Nucl Med 41: 36–44

Rinne JO, Kuikka JT, Bergström KA, Rinne UK (1995) Striatal dopamine transporter in different disability stages of Parkinson's disease studied with [123I]β-CIT SPECT. Parkinsonism Rel Dis 1: 47–51

Seibyl JP, Wallace E, Smith EO, Stabin M, Baldwin RM, Zoghbi SS, Zea-Ponce Y, Gao Y, Zhang WY, Neumeyer JL, Zubal IG, Charney DS, Hoffer PB, Innis RB (1994) Whole-body biodistribution, radiation absorbed dose and brain SPECT imaging with Iodine-123-β-CIT in healthy human subjects. J Nucl Med 35: 764–770

Suarez-Roca H, Cubeddu LX (2002) The selective serotonin reuptake inhibitor citalopram induces the storage of serotonin in catecholaminergic terminals. J Pharmacol Exp Ther 302: 174–179

Torres GE, Gainetdinov RR, Caron MG (2003) Plasma membrane monoamine transporters: Structure, regulation and function. Nat Rev Neurosci 4: 13–25

Zhou FC, Lesch K-P, Murphy DL (2002) Serotonin uptake into dopamine neurons via dopamine transporters: a compensatory alternative. Brain Res 942: 109–119

Authors' address: A. Heinz, MD, Department of Psychiatry, Charité – University Medicine Berlin, Charité Campus Mitte, Schumannstrasse 20/21, D-10117 Berlin, Germany, e-mail: andreas.heinz@charite.de

J Neural Transm (2004) [Suppl] 68: 39–51

Differential executive control impairments in early Parkinson's disease

J. Uekermann[1], I. Daum[1], M. Bielawski[1], S. Muhlack[2], S. Peters[2], H. Przuntek[2], and T. Mueller[2]

[1] Institute of Cognitive Neuroscience, and
[2] Department of Neurology, St. Josef Hospital, Ruhr-University, Bochum, Germany

Summary. Investigations concerning cognitive functions in early PD have revealed memory and executive function deficits related to dysfunction of fronto-striatal circuitry. Despite the range of data base, many previous investigations are limited because of methodological questions and inconsistencies. Thus the pattern of executive function impairments in early PD is far from being established. In the present investigation, twenty PD patients in early stages of the disease were compared to control subjects on a comprehensive neuropsychological test battery, aiming to explore their cognitive profile across a range of executive subcomponents. Results revealed impairments with respect to initiation, reasoning and planning. In summary, the present investigation shows that PD is associated with a differential executive impairment pattern which is (partly) related to disease characteristics and affective variables.

Introduction

Parkinson's Disease (PD) is frequently accompanied by a broad range of memory and executive deficits related to dysfunction of fronto-striatal circuitry (Taylor et al., 1986; Stocchi and Brusa, 2000). The impairments tend to deteriorate with the progression of the disease (Owen et al., 1992), and are presumably related to dopaminergic dysfunction (Fournet et al., 2000), although changes of other neurotransmitter systems may also be involved (Dubois et al., 1990; Aarsland et al., 2002).

Despite the broad data base, a number of inconsistencies and limitations need to be considered with respect to executive dysfunction in PD. The first controversity refers to the question of an executive deficit in PD per se. In an early investigation by Brown and Marsden (1991), PD patients showed impairments in a dual task paradigm, which was interpreted in terms of a deficit in attentional resource allocation. This finding was supported by a later study by Dalrymple-Alford et al. (1994) who reported a significant decline in PD patients' performance on a random pursuit task when digit spans had to be concomitantly recalled. Fournet et al. (1996), on the other hand, concluded that the "central executive seems to work."

In addition to the discussion about the presence of a deficit as such, the pattern of executive function impairments is far from being established. Executive functions can be defined as a higher-order cognitive construct involved in the self- regulation of goal-directed behavior (Luria, 1980; Milner et al., 1985) and include a range of different subcomponents such as planning, problem solving, initiation, inhibition and working-memory. One of the best known subdivision has been described by Smith and Jonides (1999) who distinguish between inhibition, coding, planning, task management and monitoring.

Previous studies of the neuropsychological impairments in PD have largely focused on single subcomponents such as working memory and planning (Bradley et al., 1989; Owen et al., 1995; Lewis et al., 2003) and only few attempts have been made to investigate the impairment pattern in early PD across the executive domains. In these investigations, attentional set shifting problems have often been reported, which are even present before the onset of medication (Sahakian and Owen, 1992) and are ameliorated by levodopa (Lange et al., 1992). In addition, early PD patients show planning deficits, which are reflected by prolonged thinking times but not reduced accuracy scores whereas patients in more advanced stages of the disease and patients with lesions to the frontal lobes exhibit an impairment in solution accuracy (Owen et al., 1992, 1995; Robbins et al., 1994).

Pillon et al. (1995) administered a comprehensive battery of executive tests including verbal fluency, the Wisconsin Card Sorting Test, graphic series and the Stroop test. Since the aim of this study was a comparison of PD patients with other patient groups, healthy controls were not included. In three recent investigations (Rinne et al., 2000; Vingerhoets et al., 2003; Lee et al., 2003) a range of different executive functions such as initiation, inhibition, attention and set shifting have been assessed. The patients of these investigations were, however, in advanced stages of the disease with a mean disease duration of 14,5 years (Rinne et al., 2000), 12,7 (Vingerhoets et al., 2003) and 7 years respectively (Lee et al., 2003). Executive impairments in more advanced stages of the disease have also been investigated by Robbins et al. (1994), Meco et al. (1996), Van Spaendonck et al. (1996) and Monza et al. (1998). Deficits were generally observed across these studies; it is, however, unclear, whether they are due to severity of the disease and influenced by motor disability and a range of other disease-related factors. Neuropathological dysfunctions in early PD are mainly restricted to the nigrostriatal dopaminergic pathway (Dubois and Pillon, 1997) whereas additional regions are affected in later stages of the disease. An increase in severity and a range of cognitive impairments is observed with progression of the disease (Owen et al., 1992). Deficits in early stages of the disease are presumably due to subcortical changes. This is in accordance with activation studies reporting decreased flurodopa uptake in caudate nucleus related to executive impairments (Bruck et al., 2001; Rinne et al., 2000).

As regards to executive subcomponents, the distinction between spontaneous and reactive flexibility is of particular relevance (Eslinger and Grattan, 1993). In spontaneous flexibility tasks, subjects are asked to generate various solutions, whereas reactive flexibility involves the ability to shift cognition and behaviour according to particular demands. Based on their lesion study Eslinger and Grattan (1993) concluded that spontaneous flexibility is only impaired after lesions to the prefrontal cortex. Reactive flexibility, on the other hand may be impaired due to damage of either the prefrontal cortex or basal ganglia. PD patients suffer from reactive flexibility impairments as reflected by deficits on the Wisonsin Card Sorting Test, Trail Making Test and Stroop Test (Meco et al., 1996; Monza et al., 1998; Dujardin et al., 2003). Spontaneous flexibility on the other hand should be unaffected. This assumption, however, is contrasted by the findings of Dalrymple-Alford et al. (1994) and Rogers et al. (1998) who reported lower verbal fluency scores in PD patients.

An important open issue concerns the relationship between cognitive performance and clinical symptoms. Motor disability and rigidity in particular have been reported to be related to perseveration or cognitive shifting problems (Alevriadou et al., 1999; Van Spandendonck et al., 1996), other studies did not yield significant correlations between executive and motor dysfunction (Cooper et al., 1991; Graham and Sagar, 1999).

With respect to methodological concerns, PD patients in early stages of the disease are more suitable for the study of the issues in question, since motor status should not adversely affect cognitive assessment to a major degree. Patients with dementia should be excluded, in particular within the light of the discussion whether demented PD patients represent a separate clinical entity. (Dubois et al., 1991). Similarly, the potential effect of depression on cognitive functions needs to be considered, since several studies have shown that depression may exacerbate cognitive dysfunctions in PD (Starkstein et al., 1989; Kuzis et al., 1997; Tröster et al., 1995; Uekermann et al., 2003). From the clinical point of view, ecological valid measures of executive function should be used as well as self-assessment measures of behavioral problems in order to allow an estimate of the potential impact of cognitive problems on everyday life of PD patients.

The present study aimed to assess the complex interplay of motor and cognitive impairment in early PD, focusing on the executive function system and fronto-striatal circuitry. Twenty PD patients in early stages of the disease were compared with healthy controls on a comprehensive neuropsychological battery including a range of executive function components assessed by ecologically valid procedures (Wilson et al., 1996). In addition, behavioral problems, depression, reaction times and other aspects of motor performance were considered.

Subjects and methods

Subjects

Two groups partcipated in the investigation. The first group (PD) comprised 20 PD patients (12 f, 8 m) and the second group (HC) consisted of 20 healthy controls (11 f, 9 m). The two groups were matched for age and an estimate of general intellectual function as assessed by the picture completion and similarities subtests of the Wechsler Adult Intelligence Scales (Wechsler, 1981). Mean age was 55.9 (SD = 10.3) in the PD and 53.2 (SD = 9.5) in the HC group. The mean scores on the Similarities subtest was 24.1 (SD = 4.6) in the PD and 25.2 (SD = 5.2) in the

HC group; the mean scores for Picture Completion were 13.1 (SD = 2.4) and 13.4 (SD = 2.6), respectively. All HC patients were right-handed, the PD group comprised 19 right-handed and 1 left-handed subject. The two groups did not differ significantly on age or measures of general intellectual abilities (all $p > .39$).

The Mini Mental State Examination (MMSE, Folstein et al., 1975) was administered to exclude patients with dementia. All PD patients scored above 25, all NC patients scored above 26. Clinical diagnosis of idiopathic Parkinson's Disease and assessments of disease severity were made by a neurologist who was blind to patients' neuropsychological status. Measurements of clinical and affective variables included the Hoehn and Yahr Scale (Hoehn and Yahr, 1967), the Unified Parkinson's Disease Rating Scale (Frahn et al., 1987) and the Beck Depression Inventory (Beck, 1987). Mean symptom duration in the PD group was 4.6 years (SD = 3.0), mean Hoehn and Yahr score was 1.75 (SD = 0.4, range 1–2). Mean UPDRS rating was 17.2 (SD = 7.8). The mean BDI score was 8.0 (SD = 5.7) in the PD group and 4.6 (SD = 3.4) in the HC group. The two groups differed significantly on this measure ($p = .03$).

Medication of the patients consisted of levodopa/decarboxylase inhibitor and/or dopamine agonists. Patients with on/off fluctuations were excluded. Healthy controls were recruited by advertisement and were excluded if they had a history of psychiatric or neurologic illness, head trauma, substance abuse or were taking of medication with central nervous side effects.

Measures

After having given written consent to participate in the investigation, subjects completed the following assessment which centered on a wide range of executive components such as divergent thinking, planning and problem solving, reasoning, inhibition and self-report behavioral problems in daily life.

Assessment

Reaction times and motor performance

Reaction times and motor performance were measuerd by a visual and an auditive task (Schuhfried GmbH, Austria). In both tasks, subjects have to put their index finger on a start button. They are instructed that after the presentation of a yellow light on a 31 cm × 42 cm rectangular surface, they have to switch off the light as quickly as possible by moving his/her finger from the central start button to the response button. The auditive tasks includes the presentation of a tone instead of the light. Analysis includes reaction times (RTs, time from stimulus onset to the release of start button) and motor times (MT, time from release of the start button to pressing the response button).

In addition, the tapping rate of the right and left hand within a time interval of 32 sec were measured using a contact board (3 cm × 3 cm) with a contact pencil. Tapping rate was measured by means of a computerized device.

Motor performance was further assessed by a peg insertion task. In this task, subjects have to individually transfer 25 pegs (diameter 2.5 mm, length 5 cm) as quickly as possible from a rack into one of 25 holes (diameter 2.8 mm) of a computer-based contact board. Analysis included the time interval between inserting of the first and last pin, separately for the right and left hand.

Memory

Short-term and working memory. To assess short-term memory, the digit span subtests of the Wechsler Adult Intelligence Scales (Wechsler, 1981) were administered. In the forward condition, sequences of numbers have to be repeated. In the backwards condition, subjects are instructed to reproduce series of numbers backwards.

Executive function

Initiation. Initiation was measured by three verbal fluency subtests. The semantic condition requires the production of as many names of countries as possible within 1 min. In the phonemic condition, subjects have to produce nouns beginning with the letter "b" within 1 min. In the third subtest, subjects are instructed to alternately produce male first names and vegetables within 1 min. Dependent variables are the number of correct exemplars.

Planning and problem solving. Three standard tests were administered to assess planning and problem solving abilities. The key search subtest of the Behavioral Assessment of the Dysexecutive Syndrome testbattery (BADS; Wilson et al., 1996) includes a A4-sized sheet with 100 mm square in the middle and a small black dot 50 mm below it. The subjects are asked to imagine that the paper sheet represents a large field, in which they have lost a key. They are asked to draw the route they would walk to find this key with the black dot as starting point. A score reflecting the efficiency and quality of search is calculated according to the BADS manual. In addition, time to completion is measured.

Planning was further assessed by the Modified Six Elements Test (SET) of the BADS (Wilson, 1996). Subjects are instructed to carry out three tasks (dictation, arithmetic and picture naming), each consisting of two subtests, A and B. Subjects are asked to complete at least part of each test within ten minutes. They are not allowed to do parts A and B of a test consecutively. Analyses is based on the number of rule breaks and a total score reflecting overall performance.

In addition, the Zoo Map test of the BADS (Wilson et al., 1996) was administered. This task includes the administration of a zoo map and subjects are asked to indicate the route they would choose to visit a number of designated locations. Individual paths can only be used once, starting point (zoo entrance) and end point (picnic area) are also fixed. In the second, low demand condition, subjects are instructed to simply follow instructions to achieve an error-free performance. The maximum score is 8.

Reasoning. To assess reasoning, all subjects completed the temporal judgement test of the BADS (Wilson et al., 1996). This test consists of four questions concerning the duration of events (e.g. how long does it take to blow up a party balloon?). Subjects are instructed to make a good guess. The number of correct responses was analyzed.

Reasoning was further assessed by a German version of the Cognitive Estimates Test (CET, Shallice and Evans, 1978; Daum, unpublished). The CET consists of 15 questions the answers to which can only be inferred from general knowledge (e.g. "How old is the oldest person of Germany?"). The answers are rated on a three point scale (0–3) reflecting the deviation from a realistic estimate. Higher scores are associated with poorer performance.

Inhibition. Response suppression was assessd by a German adaptation of the Hayling Test (Burgess and Shallice, 1997). The Hayling Test includes two subtests. In the first subtest, 15 incomplete sentences are read aloud and must be completed by a word which represents a meaningful completion. In the second subtest, 15 other sentences have to be completed by a word that does not fit into the semantic context of the sentence. Three scores are calculated, with higher scores indicating better performance. Score A and Score B refer to the time subjects need to complete the two subtasks. The third score (C) is calculated on the basis of the number of errors in the second subtest, indicating inhibition. In addition, a general score is derived reflecting overall performance.

Inhibition was further assessed by the rule shift cards test of the BADS (Wilson et al., 1996). In the first condition subjects have to respond "yes" to a red card and "no" to a black card. In the second condition, subjects are instructed to say "yes" if the previous card was of the same colour as the card that has just been turned over. RTs, errors as well as an overall score are assessed.

Self-report behavioral problems. Behavioral problems were assessed by the Dysexecutive Questionnaire (DEX) of the BADS (Wilson et al., 1996). The DEX consists of 20 items refering to behavioral problems frequently observed in association with executive dysfunction. One version is completed by the subject and a second by a relative or carer. Each item is rated on a 5-point Likert Scale (ranging from "never" to "very often"). Analyses includes the separate scores as well as difference scores of the two versions.

Analyses

For statistical analyses, t-tests for independent groups were performed. The association of disease variables and executive measures were assessed by correlational analyses. Since both groups significantly differed with respect to depression, ANCOVAs with the BDI score as covariate were also performed.

Results

Reaction times and motor performance

Results for RT and other motor measures are shown in Table 1.

Analysis did not yield significant group differences for RTs, MTs and tapping rates (all $p > .23$). Group comparisons for the pegboard task showed significant higher scores of PD patients for both hands (both $p < .01$).

Short-term and working memory

The PD patients scored 6.8 (SD = 2.0) on digit span forward reproduction and 6.2 (SD = 1.3) on backward reproduction, the respective scores for HC subjects were 6.7 (SD = 1.3) and 5.9 (SD = 1.7). Group comparisons did not yield significant differences ($p > .47$).

Executive function

Initiation. The number of exemplars generated in the three verbal fluency subtests as well as the results for other executive tasks are presented in Table 2.

In the semantic and alternate conditions, no significant group differences emerged (both $p > .13$). Analysis of the phonemic condition indicated significantly reduced fluency in PD patients ($t(38) = -2{,}77$; $p = .01$).

Planning and problem solving. The results for the key search, Zoo map and Modified SET are shown in Table 2. PD patients achieved a poorer overall score on the Modified SET than HC subjects ($t(38) = -2{,}89$; $p = .01$), while the other problem solving variables did not yield significant group differences (all $p > .05$).

Table 1. Results (means and SDs) for age, clinical variables and intellectual functioning

	PD	HC
Demographic variables		
Sex	12 f, 8 m	11 f, 9 m
Age	55,9 (10,31)	53,15 (9,52)
Handedness	19 r, 1 l	20 r
Intellectual functioning		
Similarities	24,05 (4,63)	25,15 (5,17)
Picture completion	13,05 (2,42)	13,35 (2,58)
Clinical variables		
Mini Mental State Examination	29,45 (1,00)	29,30 (0,98)
Beck Depression Inventory	8,00 (5,69)	4,60 (3,41)
Symptom duration (years)	4,62 (2,96)	
Hoehn and Yahr	1,75 (0,41)	
Unified Parkisnon's Disease Rating Scale	17,20 (7,81)	

Table 2. Results (means and SDs) for attention, reaction time measures and motor performance

	PD	HC
Attention		
Digit Span forward	6,80 (2,04)	6,65 (1,31)
Digit Span backward	6,25 (1,33)	5,90 (1,65)
Reaction time and motor measures		
Reaction times (visual task) msec	327,60 (104,31)	291,75 (80,91)
Reaction times (auditive task) msec	340,15 (197,21)	285,80 (79,27)
Motor times (visual task) msec	231,20 (67,78)	236,30 (70,34)
Motor times (auditive task) msec	216,10 (56,47)	224,89 (67,55)
Finger tapping (right hand)	185,90 (28,05)	187,70 (19,71)
Finger tapping (left hand)	168,60 (20,02)	174,50 (17,67)
Pegboard (right hand)	56,65 (12,12)	47,72 (6,66)
Pegboard (left hand)	61,37 (12,04)	48,26 (6,41)

Reasoning. The performance of the two groups in the temporal judgement task as well as the Cognitive Estimates Test is shown in Table 2. PD patients showed poorer temporal judgement scores than HC subjects ($t(38) = -2{,}03$; $p = .05$), while there were no significant group difference on the CET test ($p = .63$).

Inhibition. Group means for the performance on the Hayling Test and the rule shift cards test are presented in Table 3.

Group comparisons revealed significant differences with respect to the time needed to make incongruent completions on the Hayling Test as well as the overall score and the number of errors (second condition) of the rule shifts card test (all $p < .04$).

Table 3. Results (means and SDs) for attention, verbal fluency, planning and problem solving as well as reasoning

	PD	HC
Initiation		
Semantic verbal fluency	21,75 (7,69)	25,75 (8,78)
Phonemic verbal fluency	10,05 (5,16)	13,75 (3,02)
Alternate verbal fluency	14,55 (4,14)	14,15 (3,31)
Planning and problem solving		
Key Search (quality of search)	3,45 (1,10)	3,40 (0,82)
Key Search (time in sec.)	38,15 (19,05)	27,80 (13,78)
Modified Six Elements	2,85 (1,23)	3,70 (0,47)
Modified Six Elements (errors)	1,20 (3,11)	0,20 (0,41)
Zoo map task (first condition)	4,85 (3,33)	5,70 (2,87)
Zoo map task (second condition)	8,00 (0,00)	7,85 (0,67)
Reasoning		
Temporal judgement	2,80 (0,95)	3,35 (0,75)
Cognitive Estimates Test	3,50 (2,42)	3,15 (2,21)

Table 4. Results (means and SDs) for inhibition and behavioral measures

	PD	HC
Inhibition		
Hayling (Overall performance)	3,95 (1,54)	4,35 (1,35)
Hayling A	4,45 (0,76)	4,85 (0,59)
Hayling B	5,55 (0,83)	4,95 (0,89)
Hayling C	3,40 (2,39)	4,10 (1,83)
Rule Shift Cards Test		
Overall performance	2,90 (0,91)	3,55 (0,69)
First Condition (time)	31,25 (9,87)	28,30 (11,70)
First Condition (errors)	0,50 (2,24)	0,05 (0,22)
Second Condition (time)	37,85 (15,19)	31,75 (12,08)
Second Condition (errors)	2,35 (2,64)	0,90 (1,65)
Behavioral variables		
DEX (subjective)	13,40 (7,43)	13,15 (6,86)
DEX (objective)	10,20 (8,28)	10,70 (7,51)

Self-report behavioral changes. The mean self-rating scores on the DEX questionnaire were 13.4 (SD = 7.4) for PD patients and 13.2 (SD = 6.9) for HC subjects. The mean scores for ratings by family/carers were 10.2 (SD = 8.3) for PD patients and 10.7 (SD = 7.5) for controls. The group differences were not significant ($p > .84$).

Correlational analysis. Duration of the disease correlated significantly with inhibition as assessed by the rule shift cards test of the BADS ($p = .02$). Hoehn and Yahr scores showed significant correlations with the performance on the pegboard task (left hand) ($p = .03$). In addition, the motor score (in particular akinesis) of the UPDRS was significantly associated with the overall score of the rule shift cards test and the performance on the pegboard task (all $p < .04$). Rigidity was significantly associated with the scores on the pegboard task (both $p < .04$).

Effect of depression. ANCOVAS with the BDI score as covariate did not affect the overall pattern of results. The only exception was incongruent completion of sentences in the Hayling Test, where the group difference did not reach significance when group differences in depression were taken into account.

Discussion

The results of the present investigation indicate that PD patients in early stages of the disease suffer from a broad range of mild executive deficits. Executive impairments were seen for initiation, inhibition, reasoning, planning, and problem solving. The finding of a central executive deficit is in accordance with results of other neuropsychological investigations of PD patients (Taylor et al., 1986; Stocchi and Brusa, 2000).

The most important finding is that PD appeared to have a differential effect on different subcomponents of executive function. Initiation problems were reflected in reduced phonemic fluency. Inhibition impairments were observed on two tests, and reasoning problems were mirrored in temporal judgement deficits. Planning deficits were observed when task management was required within close temporal constraints.

Attention and working memory as assessed by the digit spans were intact. In addition, PD patients did not suffer from behavioural problems in everyday life, according to both their own rating and ratings by family or carers. No significant differences could be observed for other initiation measures (semantic verbal fluency, alternate verbal fluency and the first condition of the Hayling test) as well as other planning and problem solving tasks (Key Search and Zoo map task). Reasoning problems were restricted to temporal judgement only whereas general cognitive estimation was unaffected.

In the present investigation impairments occurred on tasks with time constraints (Modified Six Elements Test, Verbal fluency and Hayling Test) or tests which involved reaction time measures (rule shift cards test). The deficits were, however, not associated with an increased error rate. The general pattern of impairment is thus consistent with previous investigations, which have reported that early medicated PD patients reveal prolonged thinking times, whereas performance accuracy is intact (Owen et al., 1992; Robbins et al., 1994).

With respect to the motor assessments and RT measures, PD patients showed impaired performance on the pegboard task, whereas reaction and motor times as well as the tapping performance was unaffected Tapping, assessed with this measure, represents a task with repetitive performance and programming of standardized movements. This requires low cognitive efforts, since the subject may develop a fixed habit tendency with a consistent, attentional saving of cognitive resources (Lalonde and Botez-Marquard, 1997). Thus individuals learn a sequence of movements, which is based on an automatic function of a cognitive set (Brown and Marsden, 1990). More automatic functioning of attentional control and selective attentional processing is relative intact in PD (Claus and Mohr, 1996).

Impaired phonemic verbal fluency, which can be interpreted as reduced efficiency of access to lexical memory stores is of particular relevance in the light of studies which have reported that lexical fluency deficits could be a predictor for the development of dementia (Levy et al., 2002). In the context of the distinction of reactive and spontaneous flexibility (Eslinger and Grattan, 1993) the finding of impaired verbal fluency indicates that not only reactive flexibility is impaired in early PD.

The finding of a time estimation deficit corroborates previous observations in PD (Artieda et al., 1992). However, these studies are generally based on the investigation of the temporal discrimination threshold. The observed deficits in the present study provide the first evidence that time estimation deficits may also apply to areas relevant for the patients' everyday behaviour.

As concerns the interplay of cognitive and clinical variables, the present results yielded significant correlations between the duration of the disease, the motor score (in particular akinesia) and the performance in the inhibition task. Our findings are thus consistent with reports of a significant association of clinical measures and cognitive variables (Alevriadou et al., 1999; Van Spaendonck et al., 1996). When this finding is interpreted within the framework of spontaneous and reactive flexibility (Eslinger and Grattan, 1993), measures of reactive flexibility, but not measure of spontaneous flexibility correlated significantly with motor scores. Thus these findings may imply that only some aspects of reactive flexibility, but not spontaneous flexibility are related to the degree of loss of dopamine in the nigrostriatal system. This interpretation would be consistent with the conclusion of Tomer et al. (2002) who observed that the severity of bradykenesia was only related to one measure of reactive, but not spontaneous flexibility. In addition, Brücke et al. (1997) have reported that the severity of bradykinesia and rigidity, but not tremor correlates with the degree of dopaminergic dysfunction in the striatum.

In addition, the results of the present investigation show that the overall impairment pattern cannot be attributed to elevated depression scores of PD patients although such associations have previously been reported (Starkstein et al., 1989; Kuzis et al., 1997; Tröster et al., 1995; Uekermann et al., 2003). The BDI scores of the PD patients of these studies were higher in comparison with the present investigation. However, despite of the relatively low BDI score in the present study a significant influence was found for response supression, which indicates that even very mild depressive symptoms may contribute to executive impairments in PD.

One limitation of the present study is the relative small sample sizes of the groups. Thus further investigations including larger sample sizes are desirable to investigate the pattern of executive deficits and the relationship between cognitive variables, clinical measures and every day life performance in early PD in more detail. In summary the present investigation shows that PD patients in early stages of the disease suffer from a broad range of (mild) executive deficits, which are (partly) related to disease characteristics and affective variables.

Acknowledgements

We thank Thames Valley Test Company (Manager: M. Wilson) for letting us use a german translation of the Hayling Test. Supported by the German Research Society (Da 259/9-1).

References

Aarsland D, Laake K, Larsen JP, Janvin C (2002) Donepezil for cognitive impairment in Parkinson's disease: a randomised controlled study. J Neurol Neurosurg Psychiatry 72(6): 708–712

Alevriadou A, Katsarou Z, Bostantjopoulou S, Kiosseoglou G, Mentenopoulos G (1999) Wisconsin Card Sorting Test variables in relation to motor symptoms in Parkinson's disease. Percept Mot Skills 89(3 Pt 1): 824–830

Artieda J, Pastor MA, Lacruz F, Obeso JA (1992) Temporal discrimination is abnormal in Parkinson's disease. Brain 115: 199–210

Beck AT (1987) Beck Depression Inventory. The Psychological Corporation, San Antonio, TX

Bradley VA, Welch JL, Dick DJ (1989) Visuospatial working memory in Parkinson's disease. J Neurol Neurosurg Psychiatry 52(11): 1228–1235

Brown RG, Marsden CD (1990) Cognitive function in Parkinson's disease: from description to theory. Trends Neurosci 13(1): 21–29

Brown RG, Marsden CD (1991) Dual task performance and processing resources in normal subjects and patients with Parkinson's disease. Brain 114(Pt 1A): 215–231

Bruck A, Portin R, Lindell A, Laihinen A, Bergman J, Haaparanta M, Solin O, Rinne JO (2001) Positron emission tomography shows that impaired frontal lobe functioning in Parkinson's disease is related to dopaminergic hypofunction in the caudate nucleus. Neurosci Lett 311(2): 81–84

Brücke T, Asenbaum S, Pirker W, Djamshidian S, Wenger S, Wober C, Muller C, Podreka I (1997) Measurement of the dopaminergic degeneration in Parkinson's disease with [123I] beta-CIT and SPECT. Correlation with clinical findings and comparison with multiple system atrophy and progressive supranuclear palsy. J Neural Transm [Suppl] 50: 9–24

Burgess PW, Shallice T (1997) The Hayling and Brixton Tests. Thames Valley Test Company, Bury St. Edmunds, England

Claus JJ, Mohr E (1996) Attentional deficits in Alzheimer's, Parkinson's, and Huntington's diseases. Acta Neurol Scand 93(5): 346–351

Cooper JA, Sagar HJ, Jordan N, Harvey NS, Sullivan EV (1991) Cognitive impairment in early, untreated Parkinson's disease and its relationship to motor disability. Brain 114(Pt 5): 2095–2122

Dalrymple-Alford JC, Kalders AS, Jones RD, Watson RW (1994) A central executive deficit in patients with Parkinson's disease. J Neurol Neurosurg Psychiatry 57(3): 360–367

Daum I (unpublished) German version of the cognitive estimates test

Dubois B, Pillon B (1997) Cognitive deficits in Parkinson's disease. J Neurol 244(1): 2–8

Dubois B, Pillon B, Lhermitte F, Agid Y (1990) Cholinergic deficiency and frontal dysfunction in Parkinson's disease. Ann Neurol 28(2): 117–121

Dubois B, Boller F, Pillon B, Agid Y (1991) Cognitive deficits in Parkinson's disease. In: Boller F, Grafman X (ed) Handbook of neuropsychology, vol 5. Elsevier, Amsterdam, pp 195–240

Dujardin K, Defebvre L, Krystkowiak P, Degreef JF, Destee A (2003) Executive function differences in multiple system atrophy and Parkinson's disease. Parkinsonism Relat Disord 9(4): 205–211

Eslinger PJ, Grattan LM (1993) Frontal lobe and frontal-striatal substrates for different forms of human cognitive flexibility. Neuropsychologia 31(1): 17–28

Folstein MF, Folstein SE, McHugh PR (1975) Mini-mental state: a practical method for grading the mental state of patients for the clinican. J Psychiatr Res 12: 189–198

Fournet N, Moreaud O, Roulin JL, Naegele B, Pellat J (1996) Working memory in medicated patients with Parkinson's disease: the central executive seems to work. J Neurol Neurosurg Psychiatry 60(3): 313–317

Fournet N, Moreaud O, Roulin JL, Naegele B, Pellat J (2000) Working memory functioning in medicated Parkinson's disease patients and the effect of withdrawal of dopaminergic medication. Neuropsychology 14(2): 247–253

Frahn S, Elton RL, members of the UPDRS Developing Committee (1987) Unified Disease Parkinson's Disease Rating Scale. In: Frahn S, Marsden CD, Calne DB, Lieberman A (eds) Recent development in Parkinson's disease. Macmillan Health Care Information, Florham Park, NJ, pp 153–163

Graham JM, Sagar HJ (1999) A data-driven approach to the study of heterogeneity in idiopathic Parkinson's disease: identification of three distinct subtypes. Mov Disord 14(1): 10–20

Hoehn MM, Yahr MD (1967) Parkinsonism: onset, progression and mortality. Neurology 17: 427–442

Kuzis G, Sabe L, Tiberti C, Leiguarda R, Starkstein SE (1997) Cognitive functions in major depression and Parkinson disease. Arch Neurol 54(8): 982–986

Lalonde R, Botez-Marquard T (1997) The neurobiological basis of movement initiation. Rev Neurosci 8(1): 35–54

Lange KW, Robbins TW, Marsden CD, James M, Owen AM, Paul GM (1992) L-dopa withdrawal in Parkinson's disease selectively impairs cognitive performance in tests sensitive to frontal lobe dysfunction. Psychopharmacology 107(2–3): 394–404

Lee C, Grossman M, Morris J, Stern MB, Hurtig HI (2003) Attentional resource and processing speed limitations during sentence processing in Parkinson's disease. Brain Lang 85(3): 347–356

Levy G, Jacobs DM, Tang MX, Cote LJ, Louis ED, Alfaro B, Mejia H, Stern Y, Marder K (2002) Memory and executive function impairment predict dementia in Parkinson's disease. Mov Disord 17(6): 1221–1226

Lewis SJ, Cools R, Robbins TW, Dove A, Barker RA, Owen AM (2003) Using executive heterogeneity to explore the nature of working memory deficits in Parkinson's disease. Neuropsychologia 41(6): 645–654

Luria AR (1980) Higher cortical functions in man. Basic Books, New York

Meco G, Gasparini M, Doricchi F (1996) Attentional functions in multiple system atrophy and Parkinson's disease. J Neurol Neurosurg Psychiatry 60(4): 393–398

Milner B, Petrides M, Smith ML (1985) Frontal lobes and the temporal organization of memory. Hum Neurobiol 4(3): 137–142

Monza D, Soliveri P, Radice D, Fetoni V, Testa D, Caffarra P, Caraceni T, Girotti F (1998) Cognitive dysfunction and impaired organization of complex motility in degenerative parkinsonian syndromes. Arch Neurol 55(3): 372–378

Owen AM, James M, Leigh PN, Summers BA, Marsden CD, Quinn NP, Lange KW, Robbins TW (1992) Fronto-striatal cognitive deficits at different stages of Parkinson's disease. Brain 115(Pt 6): 1727–1751

Owen AM, Sahakian BJ, Hodges JR, Summers BA, Polkey CE, Robbins TW (1995) Dopamine-dependent frontostriatal deficits in early Parkinson's disease. Neuropsychology 9(1): 126–140

Pillon B, Gouider-Khouja N, Deweer B, Vidailhet M, Malapani C, Dubois B, Agid Y (1995) Neuropsychological pattern of striatonigral degeneration: comparison with Parkinson's disease and progressive supranuclear palsy. J Neurol Neurosurg Psychiatry 58(2): 174–179

Rinne JO, Portin R, Ruottinen H, Nurmi E, Bergman J, Haaparanta M, Solin O (2000) Cognitive impairment and the brain dopaminergic system in Parkinson disease: [18F]fluorodopa positron emission tomographic study. Arch Neurol 57(4): 470–475

Robbins TW, James M, Owen AM, Lange KW, Lees AJ, Leigh PN, Marsden CD, Quinn NP, Summers BA (1994) Cognitive deficits in progressive supranuclear palsy, Parkinson's disease, and multiple system atrophy in tests sensitive to frontal lobe dysfunction. J Neurol Neurosurg Psychiatry 57(1): 79–88

Rogers RD, Sahakian BJ, Hodges JR, Polkey CE, Kennard C, Robbins TW (1998) Dissociating executive mechanisms of task control following frontal lobe damage and Parkinson's disease. Brain 121(Pt 5): 815–842

Sahakian BJ, Owen AM (1992) Computerized assessment in neuropsychiatry using CANTAB: discussion paper. J R Soc Med 115: 271–291

Shallice T, Evans ME (1978) The involvement of the frontal lobes in cognitive estimation. Cortex 14(2): 294–303

Smith EE, Jonides J (1999) Storage and executive processes in the frontal lobes. Science 283(5408): 1657–1661

Starkstein SE, Preziosi TJ, Berthier ML, Bolduc PL, Mayberg HS, Robinson RG (1989) Depression and cognitive impairment in Parkinson's disease. Brain 112(Pt 5): 1141–1153

Stocchi F, Brusa L (2000) Cognition and emotion in different stages and subtypes of Parkinson's disease. J Neurol 247 [Suppl] 2: 114–121

Taylor AE, Saint-Cyr JA, Lang AE (1986) Frontal lobe dysfunction in Parkinson's disease. The cortical focus of neostriatal outflow. Brain 109: 845–883

Tomer R, Fisher T, Giladi N, Aharon-Peretz J (2002) Dissociation between spontaneous and reactive flexibility in early Parkinson's disease. Neuropsychiatry Neuropsychol Behav Neurol 15(2): 106–112

Tröster AI, Paolo AM, Lyons KE, Glatt SL, Hubble JP, Koller WC (1995) The influence of depression on cognition in Parkinson's disease: a pattern of impairment distinguishable from Alzheimer's disease. Neurology 45(4): 672–676

Uekermann J, Daum I, Peters S, Wiebel B, Przuntek H, Müller T (2003) Depressed mood and executive dysfunction in early Parkinson's disease. Acta Neurol Scand 107(5): 341–348

Van Spaendonck KP, Berger HJ, Horstink MW, Buytenhuijs EL, Cools AR (1996) Executive functions and disease characteristics in Parkinson's disease. Neuropsychologia 34(7): 617–626

Vingerhoets G, Verleden S, Santens P, Miatton M, De Reuck J (2003) Predictors of cognitive impairment in advanced Parkinson's disease. J Neurol Neurosurg Psychiatry 74(6): 793–796

Wechsler D (1981) Wechsler Adult Intelligence Scale – Revised. Psychological Corporation, San Antonio, TX

Wilson BA, Alderman N, Burgess P, Emslie H, Evans JJ (1996) Behavioural Assessment of the Dysexecutive Syndrome (BADS). Thames Valley Test Company, Bury St Edmunds, England

Authors' address: J. Uekermann, Faculty of Psychology, Institute of Cognitive Neuroscience, GAFO 05/607, Ruhr-University Bochum, D-44780 Bochum, Germany, e-mail: Jennifer.Uekermann@ruhr-uni-bochum.de

J Neural Transm (2004) [Suppl] 68: 53–60

Simple movement sequences better correlate to levodopa plasma levels than complex ones

S. Muhlack, S. Konietzka, D. Woitalla, H. Przuntek, and **Th. Müller**

Department of Neurology, St. Josef Hospital, Ruhr University, Bochum, Germany

Summary. Instrumental assessment of movements with a computer based device reflect the clinical response of patients with Parkinson's disease (PD) to dopaminergic stimulation. We investigated associations between levodopa plasma levels, scored motor symptoms of upper limbs and instrumental test outcomes after dopaminergic stimulation. Clinical rating scores, test outcomes for simple and complex motion series significantly improved after oral application of 250 mg of a water soluble, fast absorbed levodopa/benserazide preparation, which induced a significant increase of levodopa in plasma during a two hour interval. There was a significant association between the computed area under the curve-values of levodopa plasma concentrations and test results for simple-, but not for complex movement sequences. Performance of complex motion series additionally ask for concomitant cognitive efforts with consecutive hypothetical involvement of extranigral non dopaminergic systems. In contrast, practice of simple movements is more directly associated to the predominantly dopamine regulated motor system.

Introduction

There are instrumental tests, like computer based performance of tapping and peg insertion, for the simple objective evaluation of altered fine motor skills in patients with Parkinson's disease [PD] (Vingerhoets et al., 1997; Müller et al., 2000a, 2003; Pal et al., 2001). The peg insertion procedure resembles to another tapping test, demanding for complex movements, since an alternate tapping with the index and the middle fingers on two keys, 20 cm distant from each other, is performed. The results, obtained with this instrumental tool, showed a close relationship to motor impairment in PD patients. Moreover this complex alternate finger tapping test is looked upon as more useful for serial longitudinal studies in PD in comparison with a simultaneously evaluated simple tapping task, which only asks for tapping with one finger on one similar key (Pal et al., 2001). The results, obtained with these various motor tests, represent unspecific diagnostic markers for the diagnosis of PD. Earlier trials with the used instrumental tasks of this present study showed a limited value of an in particular simple tapping task for the reflection of the motor response to application of antiparkinsonian compounds in contrast to the complex peg insertion procedure in treated PD patients, who were followed under standardized conditions with repeated scoring of motor symptoms and performance of both instrumental tools (Müller et al., 2002a, 2003). The design of these earlier trials did not incorporate measurement of levodopa plasma concentrations (Müller et al., 2002a; Müller and Benz, 2002b). A further crucial point was the oral application of standard release levodopa/benserazide preparations combined with a relative short 90 minutes lasting time period of follow-up assessments (Müller et al., 2002a, b). Thus administration of water soluble, faster resorbed levodopa/benserazide preparations in combination with extended assessment periods and simultaneously performed measurement of levodopa plasma levels would enable better conditions for an investigational approach on the relationship between simple and/or complex movements and dopaminergic stimulation.

Objective of the present trial was to look for associations between levodopa plasma levels, scored motor symptoms of upper limbs and instrumental assessment of motion after dopaminergic stimulation under cued conditions.

Subjects and methods

Subjects

We enrolled 13 right-handed idiopathic, previously treated PD subjects (age: [mean] 57.88 ± 9.99 [SD], [range] 39.5–70.3 years; 10 male, 3 female, Hoehn and Yahr Stage [HYS]: 1.85 ± 0.63, I–III; Schwab and England Scale: 86.92 ± 4.80, 80–90%; Becks Depression Inventory [BDI]: 3.54 ± 3.15, 0–8; Hamilton Depression Scale Score [HDI]: 2.23 ± 2.39, 0–7; Mini Mental State Examination [MMSE] 29.38 ± 1.33, 26–30; Unified Parkinson's Disease Rating Scale [UPDRS] total score: 29.23 ± 12.52, 10–50; UPDRS part I (mental behaviour): 1.08 ± 1.12; 1–3, UPDRS part II (activities of daily living) 7.15 ± 2.58; 3–10; part IV (complications of therapy) 1.54 ± 2.07; 0–8) into this study. We scored study participants with the BDI and HDI, since depression may influence motor performance (Kuhn et al., 1996). Table 1 shows patients' characteristics. All participants fulfilled clinical diagnostic criteria for idiopathic PD and did not suffer from unpredictable fluctuations (Hughes et al., 1992). PD patients were taken off medication for at least 12 hours and performed an overnight fast. We excluded individuals with cerebral parenchymal lesions or atrophy, previous exposure to drugs, which affect dopaminergic neurotransmission, i.e. neuroleptics. Participants showed no clinical signs of depression or dementia.

Clinical assessment

We scored PD patients at timepoint 0 with HYS, UPDRS, Beck's Depression Inventory (BDI) and then only with part III (motor examination) of the UPDRS at timepoint 30, 60, 90, 120 minutes after intake of 250 mg of water soluble levodopa/benserazide (Madopar LT®). The performed standardized protocol eliminated putative influencing factors, i.e. sleep deficits, day time etc. Oral application of the peripherally acting dopamine receptor blocker domperidone (Motilium®) (40 mg t.i.d.) 30 minutes before levodopa intake should reduce onset and/or intensity of side effects. We performed the instrumental tests before (time point: 0 = baseline), 30, 60, 90, 120 minutes after levodopa intake in combination with simultaneous UPDRS part III scoring (Müller et al., 2000b). UPDRS raters were blinded to the results of the instrumental tasks. Participants remained in the sitting position, did not eat and drink during the study, to eliminate possible disturbing circumstances. We only considered suitable items of UPDRS part III, which reflect the motor performance of the upper extremities, for the statistical analysis and added them to the

Table 1. Patients' characteristics

	1	2	3	4	5	6	7	8	9	10	11	12	13
Sex	f	m	m	m	m	m	m	m	f	m	m	m	f
Age	63.7	54.1	69.6	50.8	39.5	63.1	46	63.9	65.9	70.3	44.2	60.11	61.2
Height	165	178	172	189	187	164	178	175	160	172	190	180	162
Weight	79	78	82	80	70	79	76	80	76	75	68	99	62
PD duration	3	5	1	0.5	3	2	7	2	3	10	4	3	5
HYS	3	3	1.5	1.5	2	2.5	1.5	1.5	2	1.5	1.5	1.5	1
S & E	80	80	90	90	80	80	90	90	90	90	90	90	90
BDI	8	4	2	7	1	8	1	3	8	0	1	3	0
HDI	7	3	1	5	1	2	0	0	6	0	2	2	0
MMSE	29	30	26	30	30	27	30	30	30	30	30	30	30
Levodopa	500	200	250		100			400	600	400			75
Levodopa retard	100	100											
Entacapone	1000	1000											
DA	Br	Br + Ro	Ro		Per	Per	Per	Br	Per		Br	Per	Ca
DA dosage	37.5	2 + 20	3		4.5	0.375	3.75	30	1.5		30	3.5	8
Selegiline									10		10	7.5	10
Amantadine	600												
UPDRS I	3	0	0	1	0	2	0	1	3	2	0	1	1
UPDRS II	10	6	6	3	9	8	9	9	10	10	4	5	4
UPDRS III	33	36	18	5	22	24	13	17	28	20	6	15	16
UPDRS IV	1	8	1	1	1	1	1	0	3	0	1	1	1

Age is given in years; height is given in cm; weight is given in kg; *m* male, *f* female, *BDI* Becks Depression Inventory Score; *HDI* Hamilton Depression Scale Score; *HYS* Hoehn and Yahr Stage; *MMSE* Mini Mental State Examination Score; *S & E* Schwab and England Scale Score; *UPDRS* Unified Parkinson's Disease Rating Scale; *UPDRS I* UPDRS part I (mental behaviour), *UPDRS II* UPDRS part II (activities of daily living); *UPDRS III* UPDRS part III (motor examination); *UPDRS IV* UPDRS part IV (complications of therapy), dosages of drugs represent the daily intake in mg; *DA* dopamine agonist; *Br* bromocriptine; *Per* pergolide; *Ro* Ropinirole; *Ca* cabergoline

UPDRS arm score. We did not differentiate between the right and left side, respectively akinesia, rigidity, tremor, in order to reduce the amount of calculations (selection of items according to Müller et al., 1997).

Peg insertion

In order to execute the peg insertion procedure we instructed subjects to transfer 25 pegs (diameter 2.5 mm, length 5 cm) from a rack into one of 25 holes (diameter 2.8 mm) in a computer-based contact board individually and as quickly as possible. The distance between rack and appropriate holes was exactly 32 cm. The board was positioned in the center and the task was carried out on each side. When transferring each peg from rack to hole, elbows were allowed to be in contact with the table. We measured the time interval between inserting of the first and the last pin initially with the right – and then the left hand. We assessed the time period for this task by a computer with an accuracy <100 ms (Müller et al., 2000a).

Tapping

In order to execute the tapping test we instructed the individuals to tap as quickly as possible on a contact board (3 cm × 3 cm) with a contact pencil for a period of 32 seconds after the initial flash of a yellow stimulus light. We did not control for peak height reached by the pencil. The board was positioned in the center and the task was carried out on each side. When performing the task, elbows were allowed to be in contact with the table. We registered the number of contacts by means of a computerized device. First we measured the rate of tapping with the right and then with the left hand (Müller et al., 2000a). In order to simplify the statistical analysis and reduce the amount of calculation, we added the results of both hands to a sum score for each motor test. Since we want to minimize learning and training effects, all PD patients were asked to practice with both tests for one minute on the day before the start of the study.

Blood samples

10 ml venous blood samples for estimation of levodopa plasma levels were taken from an antecubital vein through an indwelling cannula kept patent by an infusion of heparin in saline solution (10 U/ml). We performed venous puncture 20 minutes before the baseline investigation, to enable stable conditions in particular for the following consecutive performance of the instrumental tests. Then blood samples were taken every motor assessment time period. Blood (3 ml) was drawn with a separate syringe and discarded before each 10 ml specimen. Blood specimens were placed in EDTA-test tubes containing 100 µl of 0.5% sodium disulfite solution. The plasma obtained from rapid centrifugation was immediately frozen at −80°C until analysis within 14 days. We used reversed-phase high performance liquid chromatography in combination with electrochemical detection for the estimation of levodopa in plasma (Müller et al., 2000c).

Statistical analysis

Data showed a normal distribution according to the Kolmogorow-Smirnow test. As a result, we only performed parametric tests. We used ANCOVA with repeated measures design including UPDRS I-, II-, IV, BDI-score, Hamilton Depression score, size, sex, body weight, MMSE score, and age as covariates (Kuhn et al., 1996; Müller et al., 2000c). Then we performed a post hoc analysis with planned comparisons of data of both instrumental tasks and UPDRS III scores of both arms at baseline versus measurements after 30, 60, 90 and 120 minutes. Since we did not additionally compute results for each side separately, we reduced amount of data and calculations. The total area under the curve (AUC) value for the results of instrumental tests, UPDRS scores and levodopa plasma levels was calculated using the linear trapezoidal rule. In order to compare the changes in both instrumental tests after levodopa administration with the t-test for dependent samples, we computed the mean change of improvement in % from the baseline value according to the formula (baseline – $[30+60+90+120]/4$). We adjusted the level of significance to $p=0.017$, since we compared respectively correlated values of three variables (UPDRS scores,

Table 2. Clinical data with UPDRS motor scores of the upper extremities and levodopa plasma levels

		Baseline	30	60	90	120	AUC
Peg insertion	mean ± SD	133.62 ± 24.53	120.75 ± 20.48	118.87 ± 24.03	115.36 ± 22.50	117.17 ± 24.02	14411.16 ± 2587.35
	range	94.5–187.5	91.79–162.28	91.84–176.74	87.32–161.02	89.39–162.37	11309.10–20249.25
	p		**	***	***	***	
Tapping	mean ± SD	328.69 ± 42.54	344.31 ± 38.59	347.46 ± 39.89	349.77 ± 50.95	349.69 ± 43.35	41421.92 ± 4912.58
	range	252–396	273–417	269–415	265–423	272–417	32070–49650
	p		*	**	**	**	
UPDRS motor score	mean ± SD	11.38 ± 4.96	6.92 ± 3.15	6.08 ± 2.84	5.77 ± 2.61	6.15 ± 2.88	1504.62 ± 726.36
	range	4–19	2–14	2–11	2–10	3–11	345–3045
	p		***	***	***	***	
Levodopa plasma levels	mean ± SD	0 ± 0	777.15 ± 584.90	858.46 ± 490.92	459.46 ± 369.65	264.38 ± 284.17	66818.08 ± 32320.85
	range	0–0	0–1627	0–1744	0–1018	0–688	20940–134055
	p		***	***	**	n.s.	

* p-value <0.05; ** p-value <0.01; *** p-value <0.001; p-values represent post hoc comparisons against baseline; peg insertion results are given in seconds; tapping represents the rates within a period of 32 seconds; baseline, 30, 60, 90, 120 = timepoint 0, 30, 60, 90, 120 minutes after baseline; *AUC* computed area under the curve value; levodopa plasma levels are given in pg/ml; *SD* standard deviation; *range* minimum–maximum; *p* p-value of the post hoc analysis; significant results are bold

levodopa plasma levels, results of instrumental tests). P-values below 0.05 were discussed as significant trend.

Results

Peg insertion results significantly decreased (ANCOVA $F_{(dF\,4,\,dF\,48)} = 7.18$, $p = 0.00013$) after levodopa/benserazide intake (Table 2). Tapping rates significantly increased (ANCOVA $F_{(dF\,4,\,dF\,48)} = 3.34$, $p = 0.017$) (Table 2). There was a significant ($p = 0.48$) trend, that performance of the peg insertion task (mean change: 10.98 ± 12.01 [SD] %) better improved than the tapping task (mean change: 6.96 ± 12.18 [SD] %). UPDRS III of both arms significantly (ANCOVA $F_{(dF\,4,\,dF\,48)} = 30.50$, $p = 1.2E-12$) decreased after levodopa/benserazide application (Table 2).

A significant (ANCOVA $F_{(dF\,4,\,dF\,48)} = 12.79$, $p = 3.67E-07$) increase of levodopa plasma levels appeared after oral intake (Table 2).

There were significant correlations between the computed AUC-values of levodopa plasma levels and the AUC of tapping numbers (r [correlation coefficient] $= -0.66$, $p = 0.015$) respectively the AUC of the UPDRS results ($r = 0.66$, $p = 0.015$). AUC's of peg insertion results and tapping numbers were significantly related to each other ($r = -0.75$, $p = 0.003$) (Table 2). A significant trend appeared for the correlations between the AUC of peg insertion and the AUC of levodopa plasma levels ($r = 0.59$, $p = 0.03$) and the AUC of UPDRS arm and tapping ($r = -0.56$, $p = 0.046$). No further significant associations appeared (results not shown).

Discussion

The significant improvement of the peg insertion abilities after oral water soluble, fast release levodopa/benserazide administration under cued conditions confirm previous results in treated PD patients (Müller et al., 2002a, b, 2003). Now we also found a significant increase of the tapping rate in PD patients, whereas the tapping task was not sensitive enough to reflect the improvement of motor symptoms in earlier studies (Müller et al., 2002a, 2003). In the present study, PD patients showed a more pronounced decrease of the UPDRS score in comparison to the earlier study with a less distinct reduction of motor symptoms after oral intake of a standard release levodopa/benserazide preparation in the same dosage (Müller et al., 2002a). We assume, that the achieved, more distinct UPDRS reduction with the water soluble, fast release levodopa/benserazide formulation in the present trial is correspondingly reflected in the tapping task, which still less improved compared with the peg insertion procedure. A comparison of the efficacy and the pharmacokinetics of both levodopa formulations in the same PD patients, performed on consecutive days, is necessary to give evidence for a likely superiority of the efficacy of the fast release levodopa/benserazide preparation over the standard one.

However, the main findings of the present trial are the found relations between computed AUC's of instrumental tasks, scored motor symptoms and levodopa plasma levels. This may indicate a certain functional relationship between levodopa metabolism and motor performance (Kulisevsky, 2000). Earlier results and our performed comparison of the computed mean changes postulated a certain superiority of peg insertion in contrast to tapping (Müller et al., 2002a, 2003). However, the found correlations between the tapping task and the UPDRS motor scores, respectively the levodopa plasma concentrations, in this present trial were more robust with respect to the p-values and the correlation coefficients compared with the ones, which appeared with the peg insertion test (Müller et al., 2000a). The used tapping procedure represents a more simple instrumental task, which asks for a repetitive performance and programming of standardized movements (Müller et al., 2000a, 2003). They require low cognitive efforts, since the subject may create a fixed habit tendency with a consistent, attentional saving of cognitive resources. Therefore this simple movement sequence is hypothetically more directly regulated by the predominantly dopamine regulated striatal motor system (Lalonde and Botez-Marquard, 1997). In contrast, our used peg insertion instrumental test is more complex. It asks for additional information procedures, requires various kinds of movements in a certain sequence and additionally demands for visuospatial cognition, self-elaboration of internal strategies, sorting and planning (Braak et al., 1995; Kulisevsky et al., 1996; Lalonde et al., 1997; Kulisevsky, 2000). These various forms of complex information processing depend on different kinds of visual, cognitive and sensory inputs, all of which are altered in PD and are also regulated by non dopaminergic transmitter systems in addition to the impact of dopamine levels on cognition, located particularly in the prefrontal cortex (Pillon et al., 1989; Montgomery et al., 1991; Malapani et al., 1994; Braak et al., 1995; Kulisevsky, 2000). Therefore we assume, that the more simple, less cognitive load demanding tapping procedure better correlated to the levodopa plasma levels in contrast to the more complex peg insertion task. Nevertheless the found significant association between the AUC's of both instrumental tasks is indirectly in line with earlier results in previously untreated PD patients (Müller et al., 2000a). Limitations of our trial are the limited numbers of study participants and the medication withdrawal of only 12 hours.

In conclusion our study further establishes a relationship between levodopa plasma concentrations and motor performance in PD. Simple movement sequences, i.e. tapping, reflect this association better than complex motion series, i.e. peg insertion, due to the hypothetical additional involvement of further non dopaminergic systems with concomitant increased cognitive efforts for movement preparation and performance.

References

Braak H, Braak E, Yilmazer D, Schultz C, de Vos RA, Jansen EN (1995) Nigral and extranigral pathology in Parkinson's disease. J Neural Transm [Suppl] 46: 15–31

Hughes AJ, Daniel SE, Kilford L, Lees AJ (1992) Accuracy of clinical diagnosis of idiopathic Parkinson's disease: a clinico-pathological study of 100 cases. J Neurol Neurosurg Psychiatry 55: 181–184

Kuhn W, Heye N, Müller T, Kraus P, Klotz P, Friedrich B, Welter FL, Przuntek H (1996) The motor performance test series in Parkinson's disease is influenced by depression. J Neural Transm 103: 349–354

Kulisevsky J (2000) Role of dopamine in learning and memory: implications for the treatment of cognitive dysfunction in patients with Parkinson's disease. Drugs Aging 16: 365–379

Kulisevsky J, Avila A, Barbanoj M, Antonijoan R, Berthier ML, Gironell A (1996) Acute effects of levodopa on neuropsychological performance in stable and fluctuating Parkinson's disease patients at different levodopa plasma levels. Brain 119 (Pt 6): 2121–2132

Lalonde R, Botez-Marquard T (1997) The neurobiological basis of movement initiation. Rev Neurosci 8: 35–54

Malapani C, Pillon B, Dubois B, Agid Y (1994) Impaired simultaneous cognitive task performance in Parkinson's disease: a dopamine-related dysfunction. Neurology 44: 319–326

Montgomery EB Jr, Gorman DS, Nuessen J (1991) Motor initiation versus execution in normal and Parkinson's disease subjects. Neurology 41: 1469–1475

Müller T, Benz S (2002b) Quantification of the dopaminergic response in Parkinson's disease. Parkinsonism Rel Disord 8: 181–186

Müller T, Eising EG, Reiners C, Przuntek H, Jacob M, Kuhn W (1997) 2-[123I]-iodolisuride SPET visualizes dopaminergic loss in de-novo parkinsonian patients: is it a marker of striatal pre-synaptic degeneration? Nucl Med Commun 18: 1115–1121

Müller T, Schäfer S, Kuhn W, Przuntek H (2000a) Correlation between tapping and inserting of pegs in Parkinson's disease. Can J Neurol Sci 27: 311–315

Müller T, Woitalla D, Schulz D, Peters S, Kuhn W, Przuntek H (2000b) Tolcapone increases maximum concentration of levodopa. J Neural Transm 107: 113–119

Müller T, Woitalla D, Saft C, Kuhn W (2000c) Levodopa in plasma correlates with body weight of parkinsonian patients. Parkinsonism Rel Disord 6: 171–173

Müller T, Benz S, Przuntek H (2002a) Tapping and peg insertion after levodopa intake in treated and de novo parkinsonian patients. Can J Neurol Sci 29: 73–77

Müller T, Kuhn W, Schulte T, Przuntek H (2003) Intravenous amantadine sulphate application improves the performance of complex but not simple motor tasks in patients with Parkinson's disease. Neurosci Lett 339: 25–28

Pal PK, Lee CS, Samii A, Schulzer M, Stoessl AJ, Mak EK, Wudel J, Dobko T, Tsui JKC (2001) Alternating two finger tapping with contralateral activation is an objective measure of clinical severity in Parkinson's disease and correlates with PET [18F]-Dopa Ki. Parkinsonism Rel Disord 7: 305–309

Pillon B, Dubois B, Cusimano G, Bonnet AM, Lhermitte F, Agid Y (1989) Does cognitive impairment in Parkinson's disease result from non-dopaminergic lesions? J Neurol Neurosurg Psychiatry 52: 201–206

Vingerhoets FJ, Schulzer M, Calne DB, Snow BJ (1997) Which clinical sign of Parkinson's disease best reflects the nigrostriatal lesion? Ann Neurol 41: 58–64

Authors' address: Th. Müller, MD, Department of Neurology, St. Josef Hospital, Ruhr University Bochum, Gudrunstrasse 56, D-44791 Bochum, Germany, e-mail: thomas.mueller@ruhr-uni-bochum.de

J Neural Transm (2004) [Suppl] 68: 61–67

Clinical effects of repetitive transcranial magnetic stimulation versus acute levodopa challenge in Parkinson's disease

Ch. Börnke, T. Schulte, H. Przuntek, and **Th. Müller**

Department of Neurology, St. Josef Hospital, Ruhr University, Bochum, Germany

Summary. We studied the short-term clinical effects of 10-Hz repetitive transcranial magnetic stimulation (rTMS) of the motor hand area contralateral to the more affected limb in 12 non-fluctuating, for at least 12 hours drug free patients with Parkinson's disease (PD). We investigated the efficacy of rTMS in combination with a levodopa challenge test design under double-blind, placebo controlled conditions. Significant reductions of UPDRS III motor scores showed the treatment conditions: placebo/rTMS, levodopa/sham stimulation and levodopa/rTMS. A more detailed evaluation of arm symptoms contralateral to the stimulated brain region showed even more pronounced effects for the three conditions. There were significant differences between the mean response of the UPDRS III arm scores to the four test conditions. In conclusion our study demonstrates short-term beneficial effects of 10-Hz rTMS on motor symptoms in PD patients. A release of endogenous dopamine in subcortical structures, i.e. putamen, in response to rTMS is the most likely mechanism of action.

Introduction

Repetitive transcranial magnetic stimulation (rTMS) of the human primary motor cortex modulates cortical excitability beyond the time of stimulation. Increasing or decreasing effects are mainly determined by the frequency of magnetic stimulation (Pascual-Leone et al., 1994a; Chen et al., 1997). The clinical application of rTMS in Parkinson's disease (PD) was introduced by Pascual-Leone in 1994 (Pascual-Leone et al., 1994b). Later studies suggested beneficial effects of rTMS on parkinsonian symptoms whereas others did not report any improvement (Ghabra et al., 1999; Mally et al., 1999a, b; Siebner et al., 1999, 2000; Tergau et al., 1999; Boylan et al., 2001; de Groot et al., 2001; Shimamoto et al., 2001; Sommer et al., 2002; Okabe et al., 2003). The variation of stimulation parameters (frequency, intensity, train duration, number of trains, intertrain interval, stimulation site) and time of evaluation (during rTMS – 6 months after rTMS) from study to study may at least partly account for the controversial results. Acute drug challenge with levodopa or apomorphine is a current practice in the short-term clinical assessment of PD patients. Since longer or repeated trains of rTMS are known to improve bradykinesia in PD, we determined our stimulation parameters in order to compare the short-term effects of rTMS and levodopa challenge. Objective of our study was to investigate the efficacy of rTMS in combination with a levodopa challenge test design under double-blind placebo controlled conditions (Müller et al., 2003).

Subjects and methods

Subjects

We enrolled 12 non fluctuating patients with idiopathic PD ([mean $\pm$ standard deviation {SD}, range] age: 55.92 ± 13.09, 37–74 years; Hoehn and Yahr Scale [HYS]: 2.25 ± 0.8, 1.5–4; Unified Parkinson's Disease Rating Scale [UPDRS] mental behaviour [I]: 0.67 ± 0.78, 0–2; UPDRS activities of daily living [II]: 8.58 ± 4.74, 1–17; Beck's Depression Inventory [BDI]: 7.75 ± 4.52, 1–17; female n = 6, male = 6). They were taken off medication for at least 12 hours at the time of examination. 5 PD patients were previously untreated, the other ones received levodopa (N (number of subjects) = 5), bromocriptine (N = 3), pramipexole (N = 1); ropinirole (N = 2), lisuride (N = 1), selegiline (N = 2) and amantadine (N = 1) in combination or alone on a regular basis.

Design

All PD patients were in-patients without any contraindications to rTMS. They received the same kind of hospital food, no low protein diet. They fulfilled clinical diagnostic criteria for PD (Hughes et al., 1992). We scored patients with the UPDRS, HYS and BDI (C. B.). We applied TMS, respectively sham stimulation, under double-blind conditions. Subsequently, we administered 250 mg levodopa/benserazide (Madopar®) or placebo (Th. M.). 30, 60 and 90 minutes after drug intake patients were scored again with the UPDRS III (C. B.). Thus scoring was performed by an independent investigator. We performed all test conditions on four consecutive days in a random order. Our standardised test protocol eliminated putative influencing factors such as impaired resorption of levodopa by food intake, sleep deficits, day time etc. (Colosimo et al., 1996; Hughes et al., 1991). Application of the peripherally acting dopamine receptor blocker domperidone (Motilium®) (40 mg t.i.d.) should reduce intensity and appearance of side effects of levodopa intake even the day before the test (Parkes, 1986).

Stimulation

In order to identify subjects at risk for accidential seizures, all patients underwent conventional electroencephalography within 24 hours prior to stimulation. A figure-of-eight shaped coil and a MagLite r-25 stimulator (Dantec/Medtronic) were used. rTMS was applied over the motor hand area (M1) contralateral to the more severely affected upper limb (real stimulation) or as above with an angeled coil at 90° perpendicular to the head with one wing of the figure-of-eight in contact with the scalp (sham stimulation) (Lisanby et al., 1998). We set the stimulus intensity 10% below the motor threshold at rest. We defined motor threshold as the minimum intensity that produced five motor-evoked potentials of more than 50 μV amplitude on ten consecutive trials of single pulse TMS in the contralateral abductor digiti minimi muscle. We applied ten trains of 10 Hz rTMS with a single train duration of 10 s and an inter-train interval of 10 s. Conclusively, a total number of 1000 stimuli was administered during each stimulation session. Due to head and coil fixation constant stimulation conditions were ensured. There were no adverse side effects.

Ethics

Each subject gave written informed consent. The local ethics committee approved this study.

Statistics

All data showed a normal distribution according to the Kolmogorow-Smirnow test. As a result, we performed parametric tests. We used ANCOVA with repeated measures design including age, sex, scores of the BDI, UPDRS I, II as covariates. Then we employed a post hoc analysis with Tukeys HSD test of baseline data versus measurements after 30, 60 and 90 minutes. Additionally, we analysed the corresponding UPDRS arm scores (items 20b, 21, 22b, 23–25) contralateral to the stimulation site. We computed the mean response according to the formula (UPDRS III arm baseline – [UPDRS III arm 30 + UPDRS III arm 60 + UPDRS III arm 90]/3).

Results

Evaluation of global motor score

UPDRS III did not significantly change after placebo medication and sham stimulation (ANCOVA $F_{(dF\,3,\,dF\,33)} = 1.08$, $p = 0.037$), whereas rTMS plus placebo application significantly reduced the UPDRS motor scores (ANCOVA $F_{(dF\,3,\,dF\,33)} = 4.2$, $p = 0.012$). Levodopa administration and rTMS significantly decreased the UPDRS III scores (ANCOVA $F_{(dF\,3,\,dF\,33)} = 24.12$, $p = 1.89E-08$), whereas levodopa intake and sham stimulation also induced a significant, but less pronounced reduction of UPDRS III scores (ANCOVA $F_{(dF\,3,\,dF\,33)} = 6.79$, $p = 0.001$). There was no significant effect of the covariates (results not shown) (Table 1).

Evaluation of arm symptoms contralateral to the stimulated brain region

UPDRS III arm did not significantly change after placebo application and sham stimulation (ANCOVA $F_{(dF\,3,\,dF\,33)} = 0.51$, $p = 0.68$), whereas rTMS plus

Table 1. UPDRS III data at various points of time with p-values of Tukey HSD test between baseline and measurements 30, 60 and 90 minutes after stimulation procedure (significant results are bold)

		0	30	60	90
placebo administration + sham stimulation	mean ± SD	16.75 ± 8.04	16.54 ± 8.16	16.54 ± 8.12	16.54 ± 8.12
	range	6–32	6–32	6–32	6–32
	p		0.53	0.53	0.38
placebo administration + rTMS	mean ± SD	19.67 ± 7.61	18.58 ± 7.63	18.58 ± 7.63	18.58 ± 7.63
	range	9–32	8–34	8–34	8–34
	p		**0.009**	0.19	0.050
Levodopa intake + rTMS	mean ± SD	20 ± 7.15	16.17 ± 6.39	15.92 ± 6.08	16.17 ± 6.41
	range	9–31	7–26	7–26	7–26
	p		**0.0002**	**0.0002**	**0.0002**
Levodopa intake + sham stimulation	mean ± SD	20.25 ± 7.77	18.83 ± 7.57	18.75 ± 7.47	18.67 ± 7.29
	range	10–34	8–34	8–34	8–33
	p		**0.008**	**0.005**	**0.003**

0 = before stimulation procedure (baseline); 30 (60, 90) = 30 (60, 90) minutes after stimulation procedure; *p* p-value; *rTMS* repetitive transcranial magnetic stimulation

Table 2. UPDRS III arm data contralateral to the stimulated brain region at various points of time with p-values of Tukey HSD test between baseline and measurements 30, 60 and 90 minutes after stimulation procedure (significant results are bold)

		0	30	60	90
placebo administration + sham stimulation	mean ± SD	8.58 ± 2.43;	8.58 ± 2.61	8.42 ± 2.78	8.42 ± 2.78
	range	6–12	6–13	5–13	5–13
	p		1	0.82	0.82
placebo administration + rTMS	mean ± SD	8.58 ± 2.43	7.67 ± 2.46	7.67 ± 2.46	7.67 ± 2.46
	range	6–12	4–11	4–11	4–11
	p		**0.0003**	**0.0003**	**0.0003**
Levodopa intake + rTMS	mean ± SD	8.67 ± 2.57	6.42 ± 2.57	6 ± 2.34	6 ± 2.34
	range	6–13	3–11	3–11	3–11
	p		**0.0004**	**0.0002**	**0.0002**
Levodopa intake + sham stimulation	mean ± SD	8.58 ± 2.43	7.83 ± 1.99	7.83 ± 1.99	7.83 ± 1.99
	range	6–12	5–11	5–11	5–11
	p		**0.008**	**0.008**	**0.008**

0 = before stimulation procedure (baseline); 30 (60, 90) = 30 (60, 90) minutes after stimulation procedure; *p* p-value; *rTMS* repetitive transcranial magnetic stimulation

placebo intake significantly reduced UPDRS III arm motor scores (ANCOVA $F_{(dF\,3,\,dF\,33)} = 12.44$, $p = 1.33E-05$). Levodopa application and rTMS significantly decreased the UPDRS III arm scores (ANCOVA $F_{(dF\,3,\,dF\,33)} = 14.44$, $p = 3.53E-06$), whereas levodopa administration and sham stimulation also induced a significant, but less pronounced reduction of UPDRS III arm scores (ANCOVA $F_{(dF\,3,\,dF\,33)} = 6.06$, $p = 0.002$). There was no significant effect of covariates (results not shown) (Table 2).

Comparison of the mean UPDRS III arm scores contralateral to the stimulated brain region

The comparison of the computed mean response of the UPDRS III arm scores to the four conditions (placebo/sham stimulation: 0.11 ± 0.76 [mean ± SD]; − 1–2.33 [range]; levodopa/rTMS: 2.53 ± 1.89; 0–2.33; placebo/rTMS: 0.92 ± 0.90; 0–2; levodopa/sham stimulation: 0.75 ± 1.06; 0–3) was significant (ANCOVA $F_{(dF\,3,\,dF\,33)} = 10.32$, $p = 6.07E-05$). The post hoc analysis showed, that the comparisons of the mean response of the conditions placebo/sham stimulation versus levodopa/rTMS ($p = 0.0002$); placebo/rTMS versus levodopa/rTMS ($p = 0.006$); levodopa/sham stimulation versus levodopa/rTMS ($p = 0.002$) were significant. The comparisons of the conditions placebo/sham stimulation versus placebo/rTMS ($p = 0.3$), placebo/sham stimulation versus levodopa/sham stimulation ($p = 0.5$), levodopa/sham stimulation versus placebo/rTMS ($p = 0.9$) were not significant.

Discussion

Our present results indicate a short-term improvement of UPDRS III in PD patients after application of 10-Hz rTMS as described above and/or acute levodopa challenge. Placebo administration in combination with sham stimulation was the only condition that did not significantly improve UPDRS motor scores. Levodopa intake and rTMS induced the most pronounced reduction of UPDRS III scores. The evaluation of arm symptoms contralateral to the stimulated brain region confirmed these findings, but improvement was more distinct, which confirms previous studies (Siebner et al., 2000). Moreover, the comparison of the mean response of the UPDRS III arm scores to the four test conditions was significant. We conclude, that 10-Hz rTMS over motor areas as well as acute levodopa challenge may induce a short-term improvement of motor symptoms in PD. Since the UPDRS III decrease was more pronounced after combined application of rTMS and levodopa than after combination of sham stimulation and levodopa, we assume, that rTMS augments the beneficial effects of levodopa on motor symptoms. The contrast of our results to those of previous studies that have not reported significant beneficial effects of rTMS in PD patients may result from the use of different stimulation parameters and times of assessment after magnetic stimulation. The most favourable period of time to observe beneficial effects of rTMS in PD seems to be within one hour after magnetic stimulation (Pascual-Leone et al., 1994; Siebner et al., 1999, 2000; Sommer et al., 2002). Moreover, longer or repeated trains of rTMS with stimulation frequencies higher than 1-Hz preferably improve parkinsonian symptoms (Pascual-Leone et al., 1994; Siebner et al., 1999, 2000). Since PD patients develop significant improvements on objective measures during placebo treatment even in controlled and blinded trials attention is drawn to the application of sham stimulation and the potential induction of a powerful placebo effect by the sham procedure (Goetz et al., 2000). Only in some of the previous studies supporting clinical improvement of rTMS in PD sham stimulation with variable procedures from study to study was carried out (Siebner et al., 1999, 2000; Shimamoto et al., 2001; Sommer et al., 2002). The sham procedure employed in our study is unable to produce intracerebral current and preserves the sound of the coil discharge but does not sufficiently mimic sensation in the skin and twitch of cranial muscles. A recent study introduced an improved sham procedure using electric currents that were given with electrodes fixed on the head to mimic the sensation in real stimulation (Okabe et al., 2003). In this study, however, different stimulation parameters were determined and no short-term assessment was performed. Conclusively, their results, which suggested no add-on effects of rTMS in PD, are not comparable to ours (Okabe et al., 2003). One rationale for rTMS trials in PD is the alteration of cortical excitability affecting the presumably deficient movement-related thalamocortical drive beyond the time of stimulation. The underlying mechanism, which sustains the lasting effects of rTMS, remains to be established. A prolonged increase of excitability of corticospinal output neurons, a modulation of synaptic interaction of motorcortical excitatory and inhibitory connections due to long-term potentiation – and long-term depression – like changes and a lasting activation of the supplementary motor area via corticocortical links are discussed as potential mechanisms attributable to rTMS – induced improvement of bradykinesia in PD patients (Siebner et al., 1999). Other more recent studies demonstrated that rTMS induces regional changes in neurotransmitter and neuromodulator release as well as in gene transcription in relevant brain regions (Keck et al., 2000, 2002; Müller et al., 2000; Strafella et al., 2001). The results of rodent experiments provide in vivo evidence that rTMS of frontal brain regions has a differentiated modulatory effect on selected neurotransmitter and neuromodulator systems in distinct brain areas including an increased release of dopamine in mesolimbic and mesostriatal dopaminergic systems (Keck et al., 2000, 2002). rTMS of the left prefrontal cortex has been shown to induce the release of endogenous dopamine in the ipsilateral caudate nucleus in humans while there was no such effect after stimulation of the left occipital cortex (Strafella et al., 2001). The result of this experiment was confirmed in six healthy human subjects with the concomitant use of positron emission tomography with the radiotracer [^{11}C]raclopride (Strafella et al., 2003). This study also demonstrated, that rTMS of the left primary motor cortex caused a reduction in [^{11}C]raclopride uptake in the left putamen compared with rTMS of the left occipital cortex (Strafella et al., 2003). Thus the release of endogenous dopamine in subcortical structures in response to rTMS of the human motor cortex is proven and supports the results of our study with circumstantial evidence.

In conclusion our study demonstrates short-term beneficial effects of 10-Hz rTMS on motor symptoms in PD patients. A release of endogenous dopamine in subcortical structures, i.e. putamen, in response to rTMS is the most likely mechanism of action.

Acknowledgements

We thank U. Claußnitzer, G. Edler and B. Marchewitz for their support.

References

Boylan LS, Pullman SL, Lisanby SH, Spicknall KE, Sackheim HA (2001) Repetitive transcranial magnetic stimulation to SMA worsens complex movements in Parkinson's disease. Clin Neurophysiol 112: 259–264

Chen R, Classen J, Gerloff C, Celnik P, Wassermann EM, Hallett M, Cohen LG (1997) Depression of motor cortex excitability by low-frequency transcranial magnetic stimulation. Neurology 48: 1398–1403

Colosimo C, Merello M, Hughes AJ, Sieradzan K, Lees AJ (1996) Motor response to acute dopaminergic challenge with apomorphine and levodopa in Parkinson's disease: implications for the pathogenesis of the on-off phenomenon. J Neurol Neurosurg Psychiatry 60: 634–637

de Groot M, Hermann W, Steffen J, Wagner A, Grahmann F (2001) Contralateral and ipsilateral repetitive transcranial magnetic stimulation in patients with Parkinson's disease. Nervenarzt 72: 932–938

Ghabra MB, Hallett M, Wassermann EM (1999) Simultaneous repetitive transcranial magnetic stimulation does not speed fine movement in PD. Neurology 52: 768–770

Goetz CG, Leurgans S, Raman R, Stebbins GT (2000) Objective changes in motor function during placebo treatment in PD. Neurology 54: 710–714

Hughes AJ, Lees AJ, Stern GM (1991) Challenge tests to predict the dopaminergic response in untreated Parkinson's disease. Neurology 41: 1723–1725

Hughes AJ, Daniel SE, Kilford L, Lees AJ (1992) Accuracy of clinical diagnosis of idiopathic Parkinson's disease. J Neurol Neurosurg Psychiatry 55: 181–184

Keck ME, Sillaber I, Ebner K, Welt T, Toschi N, Kaehler ST, Singewald N, Philippu A, Elbel GK, Wotjak CT, Holsboer F, Landgraf R, Engelmann M (2000) Acute transcranial magnetic stimulation of frontal brain regions selectively modulates the release of vasopressin, biogenic amines and amino acids in the rat brain. Eur J Neurosci 12: 3713–3720

Keck ME, Welt T, Müller MB, Erhardt A, Ohl F, Toschi N, Holsboer F, Sillaber I (2002) Repetitive transcranial magnetic stimulation increases the release of dopamine in the mesolimbic and mesostriatal system. Neuropharmacology 43: 101–109

Lisanby SH, Luber BL, Schroeder C, Osman M, Finck D, Amassian VE, Arezzo J, Sackheim HA (1998) rTMS in primates: intracerebral measurement of rTMS and ECS induced voltage in vivo. Electroencephalogr Clin Neurophysiol 107: 79 P

Mally J, Stone TW (1999a) Improvement in Parkinsonian symptoms after repetitive transcranial magnetic stimulation. J Neurol Sci 162: 179–184

Mally J, Stone TW (1999b) Therapeutic and "dose-dependent" effect of repetitive microelectroshock induced by transcranial magnetic stimulation in Parkinson's disease. J Neurosci Res 57: 935–940

Müller MB, Toschi N, Kresse AE, Post A, Keck ME (2000) Long-term repetitive transcranial magnetic stimulation increases the expression of brain derived neurotrophic factor and cholecystokinin mRNA, but not neuropeptide tyrosine mRNA in specific areas of rat brain. Neuropsychopharmacology 23: 205–215

Müller T, Benz S, Börnke C, Russ H, Przuntek H (2003) Repeated rating improves value of diagnostic dopaminergic challenge tests in Parkinson's disease. J Neural Transm 110: 603–609

Okabe S, Ugawa Y, Kanazawa I, The effectiveness of rTMS in Parkinson's disease study group (2003) 0,2-Hz repetitive transcranial magnetic stimulation has no add-on effects as compared to a realistic sham stimulation in Parkinson's disease. Mov Disord 18: 382–388

Parkes JD (1986) Domperidone and Parkinson's disease. Clin Neuropharmacol 9: 517–532

Pascual-Leone A, Valls-Solé J, Wassermann EM, Hallett M (1994a) Responses to rapid-rate transcranial magnetic stimulation of the human motor cortex. Brain 117: 847–858

Pascual-Leone A, Valls-Solé J, Brasil-Neto JP, Cammarota A, Grafman J, Hallett M (1994b) Akinesia in Parkinson's disease. II. Effects of subthreshold repetitive transcranial motor cortex stimulation. Neurology 44: 892–898

Shimamoto H, Takasaki K, Shigemori M, Imaizumi T, Ayabe M, Shoji H (2001) Therapeutic effect and mechanism of repetitive transcranial magnetic stimulation in Parkinson's disease. J Neurol 248 [Suppl 3]: 48–52

Siebner HR, Mentschel C, Auer C, Conrad B (1999) Repetitive transcranial magnetic stimulation has a beneficial effect on bradykinesia in Parkinson's disease. Neuroreport 10: 589–594

Siebner HR, Rossmeier C, Mentschel C, Peinemann A, Conrad B (2000) Short-term motor improvement after sub-threshold 5-Hz repetitive transcranial magnetic stimulation of the primary motor hand area in Parkinson's disease. J Neurol Sci 178: 91–94

Sommer M, Kamm T, Tergau F, Ulm G, Paulus W (2002) Repetitive paired-pulse transcranial magnetic stimulation affects corticospinal excitability and finger tapping in Parkinson's disease. Clin Neurophysiol 113: 944–950

Strafella AP, Paus T, Barrett J, Dagher A (2001) Repetitive transcranial magnetic stimulation of the human prefrontal cortex induces dopamine release in the caudate nucleus. J Neurosci 21: RC 157: 1–4

Strafella AP, Paus T, Fraraccio M, Dagher A (2003) Striatal dopamine release induced by repetitive transcranial magnetic stimulation of the human motor cortex. Brain 126: 2609–2615

Tergau F, Wassermann EM, Paulus W, Ziemann U (1999) Lack of clinical improvement in patients with Parkinson's disease after low and high frequency repetitive transcranial magnetic stimulation. In: Paulus W, Hallett M, Rossini PM, Rothwell JC (eds) Transcranial magnetic stimulation. Elsevier, Amsterdam, pp 281–288 (EEG Suppl 51)

Authors' address: Dr. Ch. Börnke, Department of Neurology, St. Josef Hospital, Ruhr University Bochum, Gudrunstrasse 56, D-44791 Bochum, Germany, e-mail: christian.boernke@web.de

J Neural Transm (2004) [Suppl] 68: 69–78

Impact of apomorphine on BOLD signal during movement in normals

G. Schmid[1], **B. Suchan**[3], **J. Rusin**[2], **I. Daum**[3], **O. Köster**[1], **H. Przuntek**[2], **Th. Müller**[2], and **S. Peters**[2]

[1] Department of Radiology, and
[2] Department of Neurology, St. Josef Hospital, and
[3] Institute of Cognitive Neuroscience, Department of Neuropsychology, Ruhr-University, Bochum, Germany

Summary. Our group investigated modulatory effects of apomorphine on cerebral activation patterns during finger tapping movements in six healthy right-handed volunteers using an established fMRI protocol. Apomorphine application disclosed a reduction of cerebral activation to the contralateral precentral and postcentral gyrus and ipsilateral cerebellum, with a prominent net reduction of BOLD signal in cerebellar areas. These findings contradict those of similar studies performed on dopaminergic function and Parkinson's disease (PD), which predominantly found augmentation of cerebral activation patterns in normal volunteers and PD patients after dopaminergic stimulation. One conceivable explanation for our singular results would be preferred binding of apomorphine to presynaptic dopaminergic receptors, leading to inhibition of endogenous dopamine release and resultant diminished dopaminergic stimulation, reflected in diminished cerebral activation patterns. These findings warrant future consideration and further investigation of possible central inhibitory effects of dopaminergic therapy in functional imaging studies of the dopaminergic system in general and PD in particular.

Introduction

A multitude of functional imaging studies, employing various imaging modalities, have been used to investigate the anatomical correlates of different cerebral functions in normal subjects and, increasingly, neurological disease states. Among the currently available imaging modalities, functional magnetic resonance imaging (fMRI) offers certain considerable advantages, first and foremost being its superior spatial resolution compared to PET and SPECT, but also its wider availability, relatively facile data acquisition, non-invasive character and independence from radioactive ligands, and, theoretically, unrestricted repeatability.

Examination of motor function is particularly amenable to fMRI, with the possibility of performing readily practicable simple or complex paradigms and a solid foundation of anatomical knowledge of involved cerebral structures. FMRI studies have disclosed different activation patterns depending upon performance of simple versus complex, self-paced versus paced, triggered versus self-initiated, real versus imagined movement patterns, mainly localising to the contralateral primary motor cortex, contralateral supplementary motor area (SMA) and, mainly bilateral, medial and lateral/dorsolateral premotor areas (PMA), contralateral anterior cingulate cortex (ACC), superior and inferior parietal areas bilaterally, bilateral basal ganglia, and ipsilateral cerebellar hemisphere (Rao et al., 1993; Humberstone et al., 1997; van Oostende et al., 1997; Wexler et al., 1997; Wildgruber et al., 1997; Moriyama et al., 1998; Reichenbach et al., 1998; Menon et al., 1998; Deiber et al., 1999; Lepage et al., 1999; Toyokura et al., 2002).

Accordingly, disease states which compromise physiological motor function also lend themselves to functional imaging examination in seeking to unravel basic disease mechanisms or possible differential diagnostic information. Extrapyramidal diseases, especially Parkinson's disease (PD), have thus been extensively examined.

fMRI studies on PD patients have consistently shown reduced rostral supplementary motor area and dorsolateral prefrontal cortex activity (Tada, 1998), coupled with overactivity of motor cortex and caudal supplementary motor area, premotor and parietal cortical motor areas (Sabatini et al., 2000), reflecting an effort of the dopamine-denervated brain to recruit parallel motor circuits to overcome the functional deficit of diseased striatocortical motor loops (Sabatini et al., 2000; Thobois et al., 2000).

PD being an effectively treatable condition, intriguing questions arise in considering the modulation of cerebral activation patterns under pharmacological interventions. Initial functional imaging studies on the impact of dopaminergic stimulation on cerebral activation patterns in PD revealed impaired mesial frontal/supplementary motor area and dorsolateral prefrontal cortex (DLPFC) activity, as well as compensatory overactivation of parallel motor circuits. PET studies demonstrated impaired activation of the supplementary motor area and contralateral putamen in PD patients in the "off" state. Activation in the supplementary motor area and dorsolateral prefrontal cortex improved significantly after apomorphine application (Jenkins et al., 1992). 133-Xe SPECT displayed similar activation patterns in the supplementary motor area and contralateral primary sensory motor cortex in PD patients on medication as well as in controls, whereas PD patients off medication lacked such activation patterns (Rascol et al., 1992), or showed significant overactivation in the ipsilateral cerebellar hemisphere (Rascol et al., 1994). Levodopa administration has been shown to lead to a relative improvement of reduced supplementary motor area activation and concomitant reduction of overactivity of primary motor and lateral premotor cortex seen in PD patients off medication (Haslinger et al., 2001). Functional deafferentiation of cortical motor areas is inherent to the pathophysiology of akinesia in PD disease; dopaminergic substitution therapy is thought to improve the functional activity of the supplementary motor area. A recent study concluded that de novo PD patients display contralateral primary motor cortex and bilateral SMA hypoactivation, which was reversible after levodopa treatment (Buhmann et al., 2003).

Little insight has been gained into the effects of dopaminergic stimulation on cerebral activity patterns during performance of motor paradigms in healthy individuals. The above-mentioned studies included healthy controls without the application of dopaminergic agents. Only two investigations, employing [^{15}O] H_2O PET, have examined the effect of dopaminergic medication on healthy subjects, using subcutaneous apomorphine, a potent non-selective dopamine agonist with affinity for D1 and D2 receptors. These studies disclosed an increase of rCBF in ACC (Grasby et al., 1993) respectively increase in ACC, DLPFC, ventral motor cortex and decrease in the retrosplenial cingulate region after injection of 10 μg/kg (Kapur et al., 1994).

To date, fMRI investigations into pharmacological effects in PD have neglected to evaluate dopaminergic effects in healthy controls. The following study investigates cerebral activity changes using fMRI in a group of normal volunteers performing a simple internally generated motor paradigm before and after subcutaneous apomorphine injection.

Material and methods

Subjects

6 right-handed subjects (3 female, 3 male) with a mean age of 27 years participated in this study. Handedness was assessed using the Oldfield Inventory (Oldfield, 1971). Handedness median was 87%. Systematic history, clinical and neurological examination revealed no significant diseases. None regularly took concomitant medications or other substances.

Ethics

All subjects gave written informed consent. The local ethics committee approved the study. The study was conducted in accordance with the guidelines determined in the Helsinki Declaration.

Task

Subjects were asked to perform internally generated, non-cued index finger-to-thumb opposition movements at their fastest possible speed. The task was trained prior to scanning. The task was performed thrice alternatively with each hand for one minutes' duration, after which 4 mg of apomorphine were injected subcutaneously via an indwelling cannula. The paradigm was then repeated after 10 minutes. Correct task performance was continually monitored by one of the investigators (SP).

Data acquisition

Images were registered using a Siemens Symphony Scanner operating at 1.5 T. 48 axial slices of 3 mm thickness were collected using an epi 2D mosaic sequence (TE = 50, repetition time TR = 100 sec, flip angle = 90, FOV = 192×192 mm). A blocked design was applied with 3 alternating activation and rest conditions. The first 5 images of each run were discarded to allow for T1 stabilisation. T1-weighted structural images were acquired (112 slices, TR = 1900, TE = 4.2, flip angle = 15, voxel size = 1.4×1×1.5) to obtain a structural three-dimensional volume.

Data analysis

The fMRI data were analysed using SPM 99 (http://www.fil.ion.ucl.ac.uk/spm/). Functional images of each subject were realigned to the first volume and normalized according to the MNI

(Montreal Neurological Institute) brain implemented in SPM99. The images were spatially smoothed using a Gaussian kernel of 8 mm full width at half maximum. Stereotactic coordinates were transformed into Talairach space (Talairach and Tournoux, 1988) for anatomical labelling using the algorithm suggested by Brett (http://www.mrc-cbu.cam.ac.uk/Imaging/mnispace.html) and described by Duncan et al. (2000).

Results

Behavioural data

Tapping frequency was analysed offline. Mean tapping frequency was calculated for each subject and condition and further analysed using a MANOVA for repeated measures including the factors side of performance (right hand/left hand) and medication (before/after apomorphine application). Only the factor side of performance yielded a significant main effect ($F_{1,5} = 13.58$; $p = 0.01$), demonstrating consistently higher general tapping frequencies during right-handed performance. Correlation between tapping frequency and handedness quotient did not reveal any significant correlation. Tapping frequency was reduced after apomorphine application on both sides without, however, reaching statistical significance (Table 1).

Cerebral activation pattern

The finger tapping conditions were contrasted with the corresponding rest condition. The statistical threshold was set to a corrected p-value of 0.05 ($T > 4.68$) at cluster level in combination with a cluster threshold of 9 voxels. Results are listed in Table 2.

Performance of the paradigm with the right hand before apomorphine revealed activation in the contralateral precentral and postcentral gyrus, ipsilateral cerebellum and contralateral SMA (Fig. 1). Left-handed performance disclosed a much more widely distributed pattern of cerebral activation, with additional significant hemodynamic response increase in the basal ganglia (contralateral globus pallidus, bilateral putamen), contralateral cerebellum, ipsilateral thalamus and contralateral superior temporal gyrus. On the whole, hemodynamic response was more pronounced in the contralateral motor cortex (Brodmann area 4) cluster during left-handed performance than during right-handed performance.

After application of apomorphine, a marked reduction in activation distribution and overall hemodynamic response could be observed. On right-handed finger tapping, activation was restricted to the contralateral precentral and

Table 1. Performance levels (tapping frequency) in the conditions right hand/left hand, pre-apomorphine/post apomorphine

	Pre apomorphine	Post apomorphine
Right side	183.6 (20.4)	179.5 (20.8)
Left side	173.6 (25.6)	168.8 (21.9)
Mean tapping frequency (±sd)		

Table 2. Anatomical labelling, cluster size, probable Brodmann Areas (BA), T – values, and stereotactic coordinates of cortical activation according to the MNI brain before and after apomorphine application

	Cluster	T	x	y	z
Right side without apomorphine					
Cerebellum anterior lobe culmen	1332	10.51	16	−52	−24
Cerebellum anterior lobe culmen		9.15	8	−56	−18
Cerebellum posterior lobe declive		6.66	14	−68	−16
Postcentral gyrus (BA 3)	752	9.21	−36	−24	48
Precentral gyrus (BA 4)		8.66	−38	−26	60
Premotor cortex (BA 6)		7.56	−30	−14	70
SMA	9	5.04	−8	−6	54
Left side without apomorphine					
Cerebellum anterior lobe culmen	2565	14.54	−12	−52	−22
Cerebellum posterior lobe declive		8.05	22	−64	−20
Cerebellum posterior lobe declive		7.61	20	−64	−30
Precentral gyrus (BA 4)	1497	9.86	40	−20	48
Precentral gyrus (BA 4)		9.80	36	−22	70
Precentral gyrus (BA 4)		9.08	34	−22	56
Putamen lentiform nucleus	566	7.15	30	−12	−4
Globus pallidus		7.13	24	−10	2
Putamen		6.45	24	2	−2
Cerebellum posterior lobe tuber	92	7.03	48	−52	−30
Cerebellum		4.78	42	−44	−22
Superior temporal gyrus (BA 22)	40	6.64	−64	−12	0
Superior temporal gyrus (BA 22)	297	6.56	62	2	6
Insula		5.92	50	10	4
Superior temporal gyrus (BA 22)		5.81	58	14	−4
Precentral gyrus (BA 4)	86	6.49	−48	−6	42
Precentral gyrus (BA 4)		(5.46)	−8	46	
Superior temporal gyrus (BA 38)	57	6.08	−58	8	−8
Putamen	134	5.93	−22	−4	2
Putamen		5.36	−28	0	6
Thalamus ventral anterior nucleus		5.15	−16	−4	12
Parietal lobe	45	5.82	−26	−44	46
SMA	34	5.81	−10	−6	52
Postcentral gyrus	72	5.75	−56	−24	18
Superior temporal gyrus (BA 42)		5.11	−66	−26	16
Lingual gyrus	38	5.73	−2	−80	−12
Lingual gyrus	25	5.68	20	−56	−4
Cerebellum tonsi	34	5.67	18	−60	−50
Cerebellum posterior lobe declive	56	5.44	−44	−60	−26
Cerebellum posterior lobe tuber		5.32	−48	−50	−28
Pons	23	5.44	8	−30	−32
Cingulate gyrus (BA 24)	28	5.39	8	−10	50
Middle temporal lobe (BA 37)	11	5.18	46	−58	−4
Superior temporal gyrus (BA 41)	14	5.13	54	−30	10
Postcentral gyrus (BA 5)	11	5.07	44	−46	62
Postcentral gyrus (BA 40)	10	5.05	52	−26	20

(*continued*)

Table 2 (*continued*)

	Cluster	T	x	y	z
Right side with apomorphine					
Postcentral gyrus	1121	13.40	−36	−24	46
Precentral gyrus (BA 4)		7.68	−30	−14	70
Precentral gyrus (BA 4)		6.74	−48	−12	44
Cerebellum anterior lobe culmen	819	10.88	16	−52	−26
Cerebellum anterior lobe declive		6.44	24	−64	−20
Premotor cortex (BA 6)	38	5.89	−50	0	30
Postcentral gyrus (BA 2)	21	5.50	56	−20	30
Postcentral gyrus	43	5.44	−54	−22	20
Postcentral gyrus	23	5.36	60	−10	20
Left side with apomorphine					
Precentral gyrus (BA 4)	896	10.50	40	−20	50
Precentral gyrus (BA 4)		8.00	34	−18	62
Precentral gyrus (BA 4)		7.01	34	−20	70
Cerebellum anterior lobe culmen	445	10.48	−10	−54	−22
Thalamus ventral lateral nucleus	19	5.41	−14	−10	10
Cerebellum posterior lobe declive	18	5.13	−42	−68	−24

x lateral distance from midline (+ right, − left), *y* anteroposterior distance from the anterior commissure (+ anterior, − posterior), *z* height relative to the anterior commissure (+ above, − below). *x*, *y*, *z* denote the position of the voxel with peak activation levels ($p < 0{,}05$)

postcentral gyrus, a small ipsilateral postcentral focus of activity, and ipsilateral cerebellum, with a marked redistribution of hemodynamic response from cerebellar to supratentorial areas (BA 4 and 6; Fig. 1). No SMA activation was

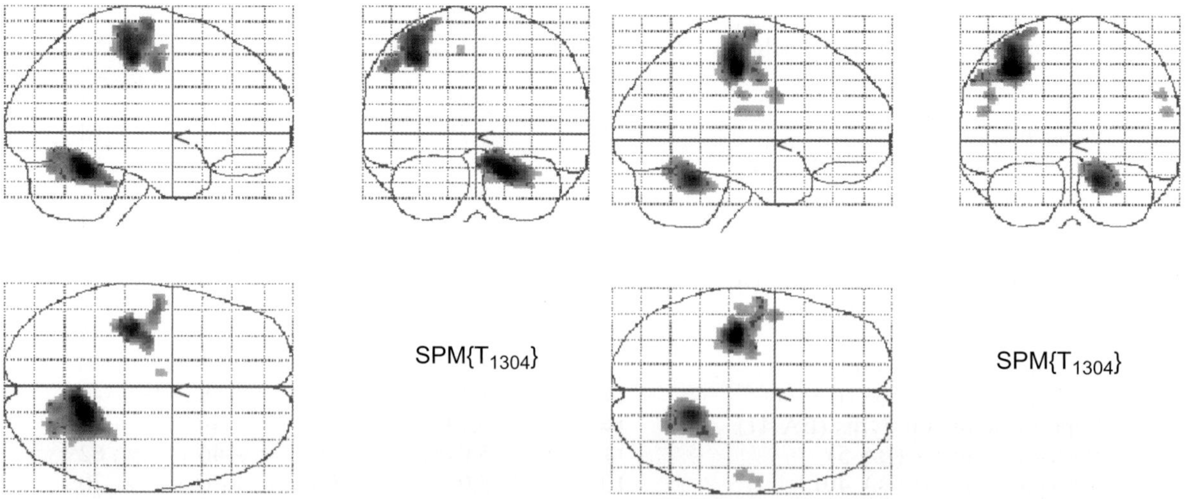

Fig. 1. Comparison of cerebral activation patterns for right-handed performance before (left) and after (right) application of apomorphine. A shift of activation is evident after apomorphine from ipsilateral cerebellar structures to contralateral motor cortex (BA 4) areas

observed post apomorphine. A much more dramatic reduction of BOLD signal distribution was observed on left-handed tapping after apomorphine, being restricted to the contralateral precentral gyrus (BA 4), ipsilateral cerebellum and thalamus. Again, a transfer of BOLD signal distribution was seen, with marked reduction of cerebellar activation.

No clinically relevant side effects, particularly sedation or nausea, were seen or reported after apomorphine application.

Discussion

Behavioural data

Tapping frequency was statistically higher on right-handed execution, which would be expected in right-handed individuals. Tapping frequencies were not statistically significantly reduced after application of apomorphine, precluding relevant sedative effects (Lal, 1988).

Cerebral activation patterns before apomorphine application

Performance of the motor paradigm disclosed differences between both hemispheres. Right-handed performance revealed a pattern of activation restricted to the contralateral pre- and postcentral gyrus, premotor cortex and SMA, and ipsilateral cerebellum, consistent with preceding fMRI and PET studies (Rao et al., 1993; van Oostende et al., 1997; Wexler et al., 1997; Moriyama et al., 1998; Reichenbach et al., 1998; Shibasaki et al., 1993; Catalan et al., 1998). Left-handed performance led to a much more extensive activation pattern, additionally encompassing the basal ganglia bilaterally, contralateral cerebellum, ipsilateral thalamus and contralateral superior temporal gyrus. Previous studies were often restricted to right-handed performance, preventing interhemispheric comparison. Rao et al. observed differences of cerebral activity between performance of complex movements with the right and left hands, but considered these laterality effects to be of uncertain significance (Rao et al., 1993). Reichenbach et al. described more bilateral activation in putamen and cerebellar regions on moving the "non-dominant thumb" (Reichenbach et al., 1998). Differing roles of both hemispheres have been described in the organization and control of complex voluntary movements (Haaland and Yeo, 1989); whether the same pertains to simple movements is uncertain. Perhaps our findings reflect increased demands on performing an identical paradigm with the non-dominant hemisphere. This could be supported by behavioural data, which, in general, demonstrate lower tapping frequencies during left hand performance. Confounding factors can be ruled out, as the motor paradigms were executed in identical conditions. The present group could be considered too small to make a valid statement on this matter, warranting studies on larger numbers of subjects. Interestingly, Peters et al. made a similar observation in PD patients under identical experimental conditions, possibly reflecting coactivation of alternate motor circuits in the dopamine-deficient central nervous systems of these patients (Peters et al., 2003); such an explanation would, naturally, not be applicable to this group.

Cerebral activation patterns after apomorphine application

Apomorphine application greatly condensed cerebral activation patterns in both conditions (right hand/left hand) to the contralateral precentral gyrus and ipsilateral cerebellar hemisphere, the contralateral postcentral gyrus on right-handed and thalamus on left-handed performance. In addition, a shift of activation from cerebellar to supratentorial motor areas could be ascertained, more pronounced on left-handed execution. This stands in contrast to studied effects of apomorphine on rCBF in animal models (McCulloch and Teasdale, 1979) and healthy volunteers (Grasby et al., 1993; Kapur et al., 1994); one study found an attenuation of rCBF in the left prefrontal cortex by apomorphine; however, this study examined cognitive tasks, not motor functions (Grasby et al., 1992).

Studies on the effect of dopaminergic medication on cerebral activation patterns in PD patients have, as a rule, shown augmentation of rCBF or BOLD signal, depending on the used imaging modality. Haslinger et al. found reduction of overactivity of primary motor and lateral premotor cortex in PD patients off medication after levodopa application as a reflection of reduced use of parallel motor circuits under dopaminergic therapy of the dopamine-denervated brain (Haslinger et al., 2001), a finding paralleled in this investigation in revealing reduced ipsilateral cerebellar activation after apomorphine application. To date, only the study of Peters et al. demonstrated consistent generalised reduction of BOLD signal in PD patients after application of apomorphine; the hypothesis for this observation was preferred binding of apomorphine to presynaptic dopaminergic receptors, leading to an inhibition of endogenous dopamine release (Peters et al., 2003). Apomorphine has been shown to cause presynaptic inhibition of endogenous dopamine release in PD patients (de la Fuentes-Fernandez et al., 2001); presynaptic dopaminergic receptors show stronger affinity to dopamine compared to their postsynaptic counterparts (Skirboll et al., 1979). These studies, however, only regarded PD patients, controls, if examined, stayed drug-free. The findings of the current study further parallel those of Peters et al. in another point – apomorphine application in PD patients shifted cerebral activity from cerebellar coactivation to exclusive activation in the contralateral pre- and postcentral gyrus, possibly a reflection of re-establishment of physiological motor loops.

In conclusion, we contend that the hypothesis of Peters et al. must be extended to normal subjects: apomorphine reduces BOLD signal mainly to the contralateral sensorimotor cortex from a more extensive pattern including the cerebellum, basal ganglia and temporal lobe during performance of a simple motor paradigm in normals as in PD patients (Peters et al., 2003). Ipsilateral cerebellar activation persists, but BOLD signal is markedly reduced. Thus, apomorphine reduces cerebral activity to the essential cerebral structures for simple motor functions. Whether this observation is dose- or time-dependent, pertains to more complex motor paradigms or other pathological states warrants further investigation. A possible mechanism for this observation is preferred presynaptic binding of apomorphine to dopaminergic receptors and reduced endogenous dopamine release, conceivably leading to altered function of cortico-subcortical loops linking the basal ganglia to premotor and prefrontal motor (Alexander et al., 1990). These observations carry potential implications for future functional imaging studies into pharmacological mechanisms in dopaminergic functions and extrapyramidal disease.

References

Alexander GE, Crutcher MD, Delong MD (1990) Basal ganglia thalamo-cortical circuits: parallel substrates for motor, oculomotor, "prefrontal" and "limbic" functions. Prog Brain Res 85: 119–146

Buhmann C, Glauche V, Stürenburg HJ, Oechsner M, Weiller C, Büchel C (2003) Pharmacologically modulated fMRI – cortical responsiveness to levodopa in drug-naive hemiparkinsonian patients. Brain 126: 451–461

Catalan MJ, Honda M, Weeks RA, Cohen LG, Hallett M (1998) The functional neuroanatomy of simple and complex sequential finger movements: a PET study. Brain 121: 253–264

Deiber MP, Honda M, Ibanez V, Sadato N, Hallett M (1999) Mesial motor areas in self-initiated versus externally triggered movements examined with fMRI: effect of movement type and rate. J Neurophysiol 81: 3065–3077

de la Fuente-Fernandez R, Lim AS, Sossi V, Holden JE, Calne DB, Ruth TJ, Stoessl AJ (2001) Apomorphine-induced changes in synaptic dopamine levels: positron emission tomography evidence for presynaptic inhibition. J Cereb Blood Flow Metab 21: 1151–1159

Duncan J, Seitz RJ, Kolodny J, Bor D, Herzog H, Ahmed A, Newell FN, Emslie H (2000) A neural basis for general intelligence. Science 289: 457–460

Grasby PM, Friston KJ, Bench CJ, Frith CD, Paulesu E, Cowen PJ, Liddle PF, Frackowiak RS, Dolan R (1992) The effect of apomorphine and buspirone on regional cerebral blood flow during the performance of a cognitive task – measuring neuromodulatory effects of psychotrophic drugs in man. Eur J Neurosci 4: 1203–1212

Grasby PM, Friston KJ, Bench CJ, Cowen PJ, Frith CD, Liddle PF, Frackowiak RS, Dolan RJ (1993) The effect of the dopamine agonist, apomorphine, on regional cerebral blood flow in normal volunteers. Psychol Med 23: 605–612

Haaland KY, Yeo RA (1989) Neuropsychological functional and neuroanatomic aspects of complex motor control. In: Bigler ED, Yeo RA, Turkheimer E (eds) Neuropsychological function and brain imaging. Plenum, New York, pp 219–243

Haslinger B, Erhard P, Kämpfe N, Boecker H, Rummeny E, Schwaiger M, Conrad B, Ceballos-Baumann AO (2001) Event-related functional magnetic resonance imaging in Parkinson's disease before and after levodopa. Brain 124: 558–570

Humberstone M, Sawle GV, Clare S, Hykin J, Coxon R, Bowtell R, Macdonald IA, Morris PG (1997) Functional magnetic resonance imaging of single motor events reveal human presupplementary motor area. Ann Neurol 42: 632–637

Jenkins IH, Fernandez W, Playford ED, Lees AJ, Frackowiak RSJ, Passingham RE, Brooks DJ (1992) Impaired activation of the supplementary motor area in Parkinson's disease is reversed when akinesia is treated with apomorphine. Ann Neurol 32: 749–757

Kapur S, Meyer J, Wilson AA, Houle S, Brown GM (1994) Activation of specific cortical regions by apomorphine: an [15O]H2O PET study in humans. Neurosci Lett 176: 21–24

Lal S (1988) Apomorphine in the evaluation of dopaminergic function in man. Prog Neuro-psychopharmacol Biol Psychiatry 12: 117–164

Lepage M, Beaudoin G, Boulet C, O'Brien I, Marcantoni W, Bourgouin P, Richer F (1999) Frontal cortex and the programming of repetitive tapping movements in man: lesion effects and functional neuroimaging. Brain Res Cogn Brain Res 8: 17–25

McCulloch J, Teasdale G (1979) Effects of apomorphine upon local cerebral blood flow. Eur J Pharmacol 55: 99–102

Menon V, Glover GH, Pfefferbaum A (1998) Differential activation of dorsal basal ganglia during externally and self paced sequences of arm movements. Neuroreport 9: 1567–1573

Moriyama T, Yamanouchi N, Kodama K, Murakami A, Okada SI, Noda S, Komatsu N, Sato T, Kusaka T, Kato K (1998) Activation of non-primary motor areas during a complex finger movement task revealed by functional magnetic resonance imaging. Psychiatry Clin Neurosci 52: 339–343

Oldfield RC (1971) The assessment and analysis of handedness: the Edinburgh inventory. Neuropsychologia 9: 97–113

Peters S, Suchan B, Rusin J, Daum I, Köster O, Przuntek H, Müller T, Schmid G (2003) Apomorphine reduces BOLD signal in fMRI during voluntary movement in Parkinsonian patients. Neuroreport 14: 809–812

Rao SM, Binder JR, Bandetti PA, Hammeke TA, Yetkin FZ, Jesmanowicz A, Lisk LM, Morris GL, Mueller WM, Estowski LD, Wong EC, Haughton VM, Hyde JS (1993) Functional magnetic resonance imaging of complex human movements. Neurology 43: 2311–2318

Rascol O, Sabatini U, Chollet F, Celsis P, Montastruc JL, Marc-Vergnes JP, Rascol A (1992) Supplementary and primary sensory motor area activity in Parkinson's disease. Arch Neurol 49: 144–148

Rascol O, Sabatini U, Chollet F, Fabre N, Senard JM, Montastruc JL, Celsis P, Marc-Vergnes JP, Rascol A (1994) Normal activation of the supplementary motor area in patients with Parkinson's disease undergoing long-term treatment with levodopa. J Neurol Neurosurg Psychiatry 5: 567–571

Reichenbach JR, Feiwell R, Kuppusamy K, Bahn M, Haacje EM (1998) Functional magnetic resonance imaging of the basal ganglia and cerebellum using a simple motor paradigm. Magn Reson Imaging 16: 281–287

Sabatini U, Boulanouar K, Fabre N, Martin F, Carel C, Colonnese C, Bozzao L, Berry I, Montastruc JL, Chollet F, Rascol O (2000) Cortical motor reorganization in akinetic patients with Parkinson's disease – a functional MRI study. Brain 123: 394–403

Shibasaki H, Sadato N, Lyshkow H, Yonekura Y, Honda M, Nagamine T, Suwazono S, Magata Y, Ikeda A, Miyazaki M, Fukuyama H, Asato R, Konishi J (1993) Both primary motor cortex and supplementary motor area play an important role in complex finger movement. Brain 116: 1387–1398

Skirboll LR, Grace AA, Bunney BS (1979) Dopamine auto- and postsynaptic receptors: electrophysiological evidence for differential sensitivity to dopamine agonists. Science 206: 89–92

Tada Y (1998) Motor association cortex activity in Parkinson's disease – a functional MRI study. Rinsho Shingeigaku 8: 729–735

Talairach J, Tournoux P (1988) Co-planar stereotaxic atlas of the human brain. Thieme, Stuttgart

Thobois S, Dominey P, Decety J, Pollak P, Gregoire MC, Broussolle E (2000) Overactivation of primary motor cortex is asymmetrical in hemiparkinsonian patients. Neuroreport 4: 785–789

Toyokura M, Muro I, Komiya T, Obara M (2002) Activation of pre-supplementary motor area (SMA) and SMA proper during unimanual and bimanual complex sequences: an analysis using functional magnetic resonance imaging. J Neuroimaging 12: 172–178

van Oostende S, van Hecke P, Sunaert S, Nuttin B, Marchal G (1997) fMRI studies of the supplementary motor area and the premotor cortex. Neuroimage 6: 181–190

Wexler BE, Fulbright RK, Lacadie CM, Skudlarski P, Kelz MB, Constable RT, Gore JC (1997) An fMRI study of the human cortical motor system response to increasing functional demands. Magn Reson Imaging 15: 385–396

Wildgruber D, Erb M, Klose U, Grodd W (1997) Sequential activation of supplementary motor area and primary motor cortex during self-paced finger movement in human evaluated by functional MRI. Neurosci Lett 227: 161–164

Authors' address: Dr. S. Peters, Department of Neurology, St. Josef Hospital, Ruhr University Bochum, Gudrunstrasse 56, D-44791 Bochum, Germany

J Neural Transm (2004) [Suppl] 68: 79–87

Worsened motor test performance following acute apomorphine injection in previously untreated patients with Parkinson's disease

Th. Müller, S. Benz, Ch. Börnke, S. Muhlack, D. Woitalla, and **H. Przuntek**

Department of Neurology, St. Josef Hospital, Ruhr University, Bochum, Germany

Summary. Instrumental tests and clinical rating assess motor disability in Parkinson's disease (PD) patients. Previous long-term dopaminergic substitution influences the behaviour following acute dopaminergic stimulation. Objective of this study was to investigate the motor response following an apomorphine application in previously untreated-, treated- and before treated PD patients, who received placebo. Outcomes of instrumental tests worsened in previously untreated-, but not in before treated PD patients after apomorphine injection and in the PD subjects under the placebo condition. Generally, rating scores of motor symptoms significantly improved after apomorphine administration, whereas placebo application showed no effects. Tolerance to sedative effects of apomorphine in treated PD patients or sensitivity of employed motor tests to presynaptic dopaminergic autoreceptor mediated inhibition of endogenous dopamine release or postsynaptic dopaminergic overstimulation with resulting decreased cognitive function in previously untreated PD patients hypothetically caused this discrepancy between outcomes of subjective clinical rating and objective motor test performance.

Introduction

Assessment of Parkinson's disease (PD) symptoms employs clinical rating scales, which suffer from interrater variability, relative insensitivity to subtle modifications or the subjective impact of the scoring examiner. Additionally, simple standardized instrumental tests, i.e. peg insertion and tapping, objectively measure altered movement in PD patients (Müller et al., 2000b, 2002b, 2003b). Peg insertion, which asks for a complex sequence of various movements, reflected the motor response to levodopa and intravenous amantadine sulphate in contrast to the tapping task, which demands for low cognitive efforts due to performance of simple, similar movements (Müller et al., 2002b, 2003b). Earlier studies, which addressed fine motor behaviour and acute antiparkinsonian drug response in PD, did not consider a putative impact of prior long-term dopaminergic drug therapy on their results. However, more recent trials found different responses of previously, untreated or treated, PD patients to various employed paradigms following an acute levodopa intake (Kulisevsky et al., 1996; Strafella et al., 2000; Müller et al., 2000a, 2002b).

Objectives of this study were to determine the effects of an acute apomorphine injection on the scored motor response and on motor test outcomes in before, untreated or treated, PD patients.

Material and methods

Subjects

We enrolled 11 (5 women, 6 men) previously untreated-, 11 (1 woman, 10 men) treated, non-fluctuating- and 11 treated (5 women, 6 men) non-fluctuating idiopathic PD patients, who only received placebo, into this study (Table 1). The before treated PD participants were taken off any antiparkinsonian medication for at least 12 hours on the investigation day. We scored the treated PD patients, who received placebo tablets, under double-blind conditions within their participation in another trial with a similar design, since we aimed to demonstrate the effect of placebo, which releases striatal endogenous dopamine, on the motor test outcomes and clinical rating (Fuente-Fernandez et al., 2001b, 2002; Fuente-Fernandez and Stoessl, 2002; Piccini et al., 2003). There were no significant differences between all three groups concerning age, Hoehn and Yahr (HYS) range, Unified Parkinson's Disease Rating Scale (UPDRS) mental behaviour (I)-, UPDRS activities of daily living (II)- and Beck's Depression Inventory (BDI) score at baseline, but UPDRS motor examination (III) rating results of the placebo treated PD patients significantly differed from the other two groups (results not shown). All participants showed no cerebral parenchymal lesion or atrophy, no signs of dementia. They were not exposed to drugs that might potentially influence the dopaminergic system.

Design

We scored PD patients with the UPDRS, HYS and BDI before subcutaneous injection of 4 mg apomorphine. Our standardized test protocol eliminated putative influencing factors, such as sleep deficits, day time etc. (Hughes et al., 1991). In order to reduce the side effects of apomorphine, we pretreated all participants with the peripherally acting dopamine receptor blocker domperidone (Motilium®) (40 mg t.i.d.) both, the day before and 30 minutes before the apomorphine application. We chose the apomorphine- and domperidone doses in accordance to the experiences and conditions of previous trials with clinical acceptable side effect outcomes (Müller et al., 2002a, 2003a). We performed both instrumental tasks and UPDRS III ratings before (time point: 0 = baseline), 30, 60 and 90 minutes after apomorphine- respectively placebo application. Raters were masked to the results.

Methods

Peg insertion

We instructed subjects to transfer 25 pegs (diameter 2.5 mm, length 5 cm) from a rack into one of 25 holes (diameter 2.8 mm) in a computer-based contact board individually and as quickly as possible. The distance between rack and appropriate holes was exactly 32 cm. The board was positioned in the middle and the task was carried out on each side. When transferring each peg from rack to hole, elbows were allowed to be in contact with the table. We measured the time interval between inserting of the first and the last pin initially with the right- and then the left hand. We assessed the time period for this task by a computer with an accuracy <100 ms (Müller et al., 2000b).

Tapping

We asked PD patients to tap as quickly as possible on a contact board (3 cm × 3 cm) with a contact pencil for a period of 32 seconds after the initial flash of a yellow stimulus light. We did not

Table 1. Demographic data of groups of PD patients

	Age	HYS	UPDRS I	UPDRS II	UPDRS III	BDI
Previously untreated	57.9 ± 9.7; 37–69	2.3 ± 0.7; I–III	0.5 ± 0.5; 0–1	7.1 ± 3.6; 3–15	26.9 ± 10.4; 14–45	5.4 ± 3.3; 0–9
Before treated	59.5 ± 6.9; 47–70	2.5 ± 0.7; I–III	1.5 ± 1.6; 0–5	10.5 ± 4.0; 1–15	31.8 ± 11.8; 17–63	7.9 ± 6.6; 0–18
Placebo condition	55.3 ± 13.5; 37–74	2.2 ± 0.8; I–III	0.7 ± 0.8; 0–2	8 ± 4.5; 1–17	15.1 ± 6.8; 6–26	8 ± 4.6; 1–17

BDI Beck's Depression Inventory; *HYS* Hoehn and Yahr Scale; all data are given as mean ± standard deviation; minimum–maximum; *UPDRS* Unified Parkinson's Disease Rating Scale, *I* mental behaviour, *II* activities of daily living, *III* motor examination; age is given in years

control for peak height reached by the pencil. The board was positioned in the middle and the task was carried out on each side. When performing the task, elbows were allowed to be in contact with the table. We registered the number of contacts by means of a computerized device. First we measured the rate of tapping with the right and then with the left hand (Müller et al., 2000b).

We allowed all participants to familiarize with both motor tasks for an interval of 60 seconds to reduce or avoid learning and training effects on test performance (Nutt et al., 2000).

Statistics

Data showed a normal distribution according to the Kolmogorow-Smirnow test. As a result, we only performed parametric tests. We used ANCOVA with repeated measures design including UPDRS I-, II-, BDI-score, body weight, sex and age as covariates (Müller et al., 2000c). Then we performed a post hoc analysis with planned comparisons of data of both instrumental tasks and UPDRS III scores of both arms (selection of items according to Müller et al., 1997) at baseline versus measurements after 30, 60 and 90 minutes. P-values below 0.05 were significant in the post hoc analysis. We calculated the mean change of the UPDRS III arm scores and the data of both instrumental tasks at the various moments according to the formula $[([30\,\text{min}] + [60\,\text{min}] + [90\,\text{min}])/3 - [0\,\text{min}] = \text{mean change}]$ for the correlation analysis with linear regression. We did not compute results of each side separately in order to reduce data and calculations. We present no comparisons to matched controls, since earlier studies already described significant differences between PD patients and healthy volunteers (Müller et al., 2000b).

Ethics

Each subject gave written informed consent. The local ethics committee approved this study.

Results

Comparisons

UPDRS III arm scores significantly (previously untreated: ANCOVA $F_{(dF\,3,\,dF\,30)} = 5.25$, $p = 0.005$; treated: ANCOVA $F_{(dF\,3,\,dF\,30)} = 10.11$, $p = 9.18E-05$) improved after apomorphine injection. Placebo administration did not significantly change UPDRS rating results (ANCOVA $F_{(dF\,3,\,dF\,30)} = 1.0$, $p = 0.4$) (Table 2).

Peg insertion outcomes significantly increased and thus worsened in previously untreated PD patients (ANCOVA $F_{(dF\,3,\,dF\,30)} = 5.7$, $p = 0.003$) in particular 30- and 60 minutes following apomorphine application. No significant change appeared in the other two groups of PD participants (treated: ANCOVA $F_{(dF\,3,\,dF\,30)} = 1.5$, $p = 0.23$; placebo: ANCOVA $F_{(dF\,3,\,dF\,30)} = 1.12$, $p = 0.36$) (Table 2).

Tapping did not significantly change (previously untreated: $F_{(dF\,3,\,dF\,30)} = 2.42$, $p = 0.09$; treated: ANCOVA $F_{(dF\,3,\,dF\,30)} = 1.12$, $p = 0.36$; placebo: ANCOVA $F_{(dF\,3,\,dF\,30)} = 0.93$, $p = 0.44$), but previously untreated PD patients showed a significant decrease and thus worsening of their tapping rates after 30 and 60 minutes according to the post hoc analysis (Table 2).

There was no significant impact of covariates.

Correlation analysis

No significant associations appeared between computed mean changes of UPDRS III arm scores and motor test outcomes in all groups of PD patients. There was no impact of age, BDI-, UPDRS I- and UPDRS II-scores on instrumental test results.

Table 2. Outcomes of peg insertion and tapping at various time points of previously untreated, before treated PD patients after apomorphine application and treated PD subjects under the placebo condition

Groups		0	30	60	90
Untreated	peg insertion	98.62 ± 20.35; 708–1263	111.47 ± 27.54; 798–1568	106.36 ± 25.06; 746–1492	102.29 ± 22.37; 726–1388
	p		**0.0005**	**0.025**	0.27
	tapping	321.1 ± 48.2; 251–406	307.5 ± 41.1; 260–394	309.1 ± 51.8; 221–386	316.6 ± 54.1; 230–395
	p		**0.026**	**0.047**	0.45
	UPDRS III (arm)	14.8 ± 4.3; 9–21	11.3 ± 4.8; 4–22	13.1 ± 4.9; 6–21	13 ± 4.1; 8–19
	p		**0.0004**	0.06	0.05
Treated	peg Insertion	114.25 ± 23.07; 833–1677	117.74 ± 31.02; 837–1890	110.13 ± 24.59; 748–1568	112.25 ± 23.27; 771–1535
	p		0.36	0.28	0.60
	tapping	297.2 ± 72.2; 139–391	291.0 ± 70.0; 136–399	300.0 ± 72.6; 148–407	344.0 ± 154.2; 152–762
	p		0.85	0.93	0.16
	UPDRS III (arm)	17.4 ± 6.3; 10–33	13.2 ± 5.2; 9–28	13.3 ± 6.4; 7–30	15.1 ± 6.0; 8–31
	p		**4.28E−05**	**5.73E−05**	**0.014**
Placebo	peg insertion	96.88 ± 24.98; 703–1633	98.67 ± 30.49; 675–1812	96.48 ± 28.11; 672–1700	95.69 ± 29.91; 695–1783
	p		0.30	0.81	0.48
	tapping	322.5 ± 41.9; 241–390	315.9 ± 39.5; 247–368	317.2 ± 37.9; 248–377	315.8 ± 37.6; 247–382
	p		0.17	0.26	0.16
	UPDRS III (arm)	11.4 ± 4.4; 6–18	11.4 ± 4.4; 6–18	10.9 ± 4.7; 6–18	10.9 ± 4.7; 6–18
	p		0.57	0.16	0.16

All data are given as mean ± standard deviation; minimum–maximum; p-values of planned comparisons between baseline and measurements 30, 60 and 90 minutes after apomorphine injection (significant results are bold); peg insertion data are given in seconds; 0 = before apomorphine or placebo application (baseline); 30 (60, 90) = 30 (60, 90) minutes after apomorphine or placebo application; *p* p-value, *UPDRS* Unified Parkinson's Disease Rating Scale; *III* motor examination

Side effects

No serious adverse events occurred in either groups. Yawning and complaints of tiredness occurred in four previously untreated and four before treated PD patients after apomorphine application.

Discussion

The main finding of our study is that performance of both instrumental tests worsened 30 respectively 60 minutes after apomorphine injection only in previously untreated PD patients but not in before treated PD patients in contrast to the scored improvement of motor symptoms. Since four PD patients in each group of the apomorphine treated PD participants complained fatigue, we assume, that the effect of an apomorphine mediated sedation does not essentially account for the outcomes of both instrumental tasks in the course of the trial.

However the missing deterioration of fine motor behaviour may result from a certain tolerance to the sedative and emetic effects of apomorphine in the before treated PD patients (Ruzicka et al., 1994). There is growing but controversial evidence for dopamine receptor mediated sedation, since PD patients often complain fatigue following intake of dopaminergic drugs and clinicians have observed yawning as aura for onset of levodopa-induced "on"-phases in PD (Goren and Friedman, 1998; Ferreira et al., 2000). Long-term dopaminergic stimulation induces tolerance to nausea and hypotension, but no study has demonstrated tolerance to fatigue, to date (Montastruc et al., 1996). Our results of worsened tapping and peg insertion abilities in previously untreated PD patients support the hypothesis, that there is greater sedation from less dopamine tolerance and therefore this may cause an arousal problem in the PD patients (Andreu et al., 1999).

A further explanation for our found discrepancy between motor test outcomes and clinical rating results focus on the binding behaviour of apomorphine to dopaminergic reuptake sites. We hypothesize, that our applied apomorphine dose preferentially binded to presynaptic dopaminergic autoreceptors. These terminals are at least six to ten times more sensitive to dopamine agonists compared to postsynaptic dopaminergic receptors (Skirboll et al., 1979). Thus apomorphine may inhibit locomotor activity by decrease of release of endogenous dopamine (Fuente-Fernandez et al., 2001a). Deterioration of tapping and peg insertion outcomes only occurred in the previously untreated PD patients, because the still existing presynaptic dopaminergic autoreceptor function mediated this inhibitory feedback regulation. Since the prior long-term dopaminergic substitution in the treated PD patients may downregulate the presynaptic endogenous dopamine synthesis due to the diminished dopaminergic autoreceptor function and additionally induce some slowly evolving postsynaptic reduced sensitivity of dopaminergic terminals, no significant change of motor test outcomes appeared in the before treated PD patients (Rioux et al., 1997; Antonini et al., 1997; Ekesbo et al., 1999). We assume that only the employed objective motor tests were sensitive to this effect of reduced presynaptic endogenous dopamine release in the previously untreated PD patients, but not the subjective clinical rating.

A further hypothesis for the worsened performance of instrumental tests in previously untreated PD patients is that there was too much dopaminergic stimulation caused by the used apomorphine dose. This may reduce cognition processing speed in the prefrontal cortex due to a disturbed balance between acetylcholine and dopamine, which results from reduced cholinergic neurotransmission in prefrontal areas (Yang and Mogenson, 1990; Prasher and Findley, 1991; Iversen, 1998; Friedman et al., 1999). Before treated PD patients were not sensitive to this dopaminergic overflow because of their previous long-term dopamine substitution.

Our earlier study showed, that acute levodopa administration does not change tapping rates in contrast to a significant reduction of peg insertion intervals in previously untreated PD patients (Müller et al., 2002b). Therefore we concluded, that this lack of effect on tapping performance resulted from the different cognitive demands of the two tests: tapping asks for automatic movements, whereas peg insertion needs more complex movements and thus more higher cognitive functions (Muller et al., 2002b). In contrast, results of our present trial showed worsened performance of both motor tests after apomorphine injection in previously untreated PD patients. In view of both studies, we now hypothesize, that all these variations or not variations may also depend on the intensities of the side effects produced by the drug at that given dose in the group used or not used to the drug. Therefore a further trial with a repeated test procedure including application of at least 1.5 mg, 3 mg and 4 mg apomorphine should assess the effect of various apomorphine doses on the motor test performance in the same PD individuals.

Nevertheless, we indirectly confirm results, which showed a different motor response duration to apomorphine and occurrence of a behavioural tolerance associated with pulsatile dopaminergic stimulation by apomorphine in treated more advanced PD patients under chronic levodopa therapy compared with previously untreated ones (Grandas and Obeso, 1989; Grandas et al., 1992). In order to confirm our findings more convincingly, a future trial is necessary in PD patients before and during long-term dopaminergic substitution, tested under cross-over double blind conditions.

In conclusion, the found discrepancy between worsened motor test performance and improvement of clinical rating suggests that additional use of standardized objective instrumental tasks may improve the validity of clinical trials, which investigate the motor benefit of antiparkinsonian compounds in particular in previously untreated PD patients. Moreover our results support considerations for a clinical differentiation of PD patients with respect to the long-term effects of dopaminergic drugs (Gordon and Reilmann, 1999).

Acknowledgements

We thank B. Marchewitz, G. Edler, and U. Claussnitzer for technical assistance.

References

Andreu N, Chale JJ, Senard JM, Thalamas C, Montastruc JL, Rascol O (1999) L-Dopa-induced sedation: a double-blind cross-over controlled study versus triazolam and placebo in healthy volunteers. Clin Neuropharmacol 22: 15–23

Antonini A, Schwarz J, Oertel WH, Pogarell O, Leenders KL (1997) Long-term changes of striatal dopamine D2 receptors in patients with Parkinson's disease: a study with positron emission tomography and [11C]raclopride. Mov Disord 12: 33–38

Ekesbo A, Rydin E, Torstenson R, Sydow O, Laengstrom B, Tedroff J (1999) Dopamine autoreceptor function is lost in advanced Parkinson's disease. Neurology 52: 120–125

Ferreira JJ, Galitzky M, Montastruc JL, Rascol O (2000) Sleep attacks and Parkinson's disease treatment [letter] [see comments]. Lancet 355: 1333–1334

Friedman JI, Temporini H, Davis KL (1999) Pharmacologic strategies for augmenting cognitive performance in schizophrenia. Biol Psychiatry 45: 1–16

Fuente-Fernandez R, Stoessl AJ (2002) The placebo effect in Parkinson's disease. Trends Neurosci 25: 302–306

Fuente-Fernandez R, Lim AS, Sossi V, Holden JE, Calne DB, Ruth TJ, Stoessl AJ (2001a) Apomorphine-induced changes in synaptic dopamine levels: positron emission tomography evidence for presynaptic inhibition. J Cereb Blood Flow Metab 21: 1151–1159

Fuente-Fernandez R, Ruth TJ, Sossi V, Schulzer M, Calne DB, Stoessl AJ (2001b) Expectation and dopamine release: mechanism of the placebo effect in Parkinson's disease. Science 293: 1164–1166

Fuente-Fernandez R, Phillips AG, Zamburlini M, Sossi V, Calne DB, Ruth TJ, Stoessl AJ (2002) Dopamine release in human ventral striatum and expectation of reward. Behav Brain Res 136: 359–363

Gordon AM, Reilmann R (1999) Getting a grasp on research: does treatment taint testing of parkinsonian patients? [letter]. Brain 122 (Pt 8): 1597–1598

Goren JL, Friedman JH (1998) Yawning as an aura for an L-dopa-induced "on" in Parkinson's disease [see comments]. Neurology 50: 823

Grandas F, Obeso JA (1989) Motor response following repeated apomorphine administration is reduced in Parkinson's disease. Clin Neuropharmacol 12: 14–22

Grandas F, Gancher ST, Rodriguez M, Lera G, Nutt JG, Obeso JA (1992) Differences in the motor response to apomorphine between untreated and fluctuating patients with Parkinson's disease. Clin Neuropharmacol 15: 13–18

Hughes AJ, Lees AJ, Stern GM (1991) Challenge tests to predict the dopaminergic response in untreated Parkinson's disease. Neurology 41: 1723–1725

Iversen SD (1998) The pharmacology of memory. CR Acad Sci III 321: 209–215

Kulisevsky J, Avila A, Barbanoj M, Antonijoan R, Berthier ML, Gironell A (1996) Acute effects of levodopa on neuropsychological performance in stable and fluctuating Parkinson's disease patients at different levodopa plasma levels. Brain 119 (Pt 6): 2121–2132

Montastruc JL, Llau ME, Senard JM, Tran MA, Rascol O, Montastruc P (1996) A study of tolerance to apomorphine. Br J Pharmacol 117: 781–786

Müller T, Eising EG, Reiners C, Przuntek H, Jacob M, Kuhn W (1997) 2-[123I]-iodolisuride SPET visualizes dopaminergic loss in de-novo parkinsonian patients: is it a marker of striatal pre-synaptic degeneration? Nucl Med Commun 18: 1115–1121

Müller T, Benz S, Przuntek H (2000a) Choice reaction time after levodopa challenge in parkinsonian patients. J Neurol Sci 181: 98–103

Müller T, Schäfer S, Kuhn W, Przuntek H (2000b) Correlation between tapping and inserting of pegs in Parkinson's disease. Can J Neurol Sci 27: 311–315

Müller T, Woitalla D, Saft C, Kuhn W (2000c) Levodopa in plasma correlates with body weight of parkinsonian patients. Parkinsonism Rel Disord 6: 171–173

Müller T, Benz S, Przuntek H (2002a) Apomorphine delays simple reaction time in Parkinsonian patients. Parkinsonism Rel Disord 8: 357–360

Müller T, Benz S, Przuntek H (2002b) Tapping and peg insertion after levodopa intake in treated and de novo parkinsonian patients. Can J Neurol Sci 29: 73–77

Müller T, Benz S, Bornke C, Russ H, Przuntek H (2003a) Repeated rating improves value of diagnostic dopaminergic challenge tests in Parkinson's disease. J Neural Transm 110: 603–609

Müller T, Kuhn W, Schulte T, Przuntek H (2003b) Intravenous amantadine sulphate application improves the performance of complex but not simple motor tasks in patients with Parkinson's disease. Neurosci Lett 339: 25–28

Nutt JG, Lea ES, Van Houten L, Schuff RA, Sexton GJ (2000) Determinants of tapping speed in normal control subjects and subjects with Parkinson's disease: differing effects of brief and continued practice. Mov Disord 15: 843–849

Piccini P, Pavese N, Brooks DJ (2003) Endogenous dopamine release after pharmacological challenges in Parkinson's disease. Ann Neurol 53: 647–653

Prasher D, Findley L (1991) Dopaminergic induced changes in cognitive and motor processing in Parkinson's disease: an electrophysiological investigation. J Neurol Neurosurg Psychiatry 54: 603–609

Rioux L, Frohna PA, Joyce JN, Schneider JS (1997) The effects of chronic levodopa treatment on pre- and postsynaptic markers of dopaminergic function in striatum of parkinsonian monkeys. Mov Disord 12: 148–158

Ruzicka E, Roth J, Spackova N, Mecir P, Jech R (1994) Apomorphine induced cognitive changes in Parkinson's disease. J Neurol Neurosurg Psychiatry 57: 998–1001

Skirboll LR, Grace AA, Bunney BS (1979) Dopamine auto- and postsynaptic receptors: electrophysiological evidence for differential sensitivity to dopamine agonists. Science 206: 80–82

Strafella AP, Valzania F, Nasetti SA, Tropeani A, Bisulli A, Santangelo M, Tassinari CA (2000) Effects of chronic levodopa and pergolide treatment on cortical excitability in patients with Parkinson's disease: a transcranial magnetic stimulation study. Clin Neurophysiol 111: 1198–1202

Yang CR, Mogenson GJ (1990) Dopaminergic modulation of cholinergic responses in rat medial prefrontal cortex: an electrophysiological study. Brain Res 524: 271–281

Authors' address: Th. Müller, MD, Department of Neurology, St. Josef Hospital, Ruhr University Bochum, Gudrunstrasse 56, D-44791 Bochum, Germany, e-mail: thomas.mueller@ruhr-uni-bochum.de

J Neural Transm (2004) [Suppl] 68: 89–95

Transdermal lisuride delivery in the treatment of Parkinson's disease

D. Woitalla[1], **T. Müller**[1], **S. Benz**[1], **R. Horowski**[2], and **H. Przuntek**[1]

[1] Department of Neurology, St. Josef Hospital, Ruhr University Bochum, and
[2] NeuroBiotec Berlin, Berlin, Germany

Summary. Transdermal delivery of dopamine agonists (DA) is a promising therapeutic concept, which aims to ameliorate frequency and intensity of motor fluctuations in patients with Parkinson's disease (PD). We treated 8 PD patients with unpredictable on-off phenomena with lisuride patches (release: 2–5 μg lisuride base/cm^2/hour in mice) in addition to their preexisting antiparkinsonian drug regime up to a period of 8 days. In order to quantify the intensity and frequency of motor fluctuations, we determined the motor changing rate (MCR), which corresponds to the patient's self rating of motor function, performed every thirty minutes, divided through the number of scored intervals minus 1. Additional lisuride patch application significantly ($p = 0.023$) improved the MCR compared to baseline. Relevant side effects were transient skin irritations in four patients. Our observational study demonstrates the safety, tolerability and efficacy of transdermal lisuride delivery in the treatment of motor complications.

Introduction

A major drawback of the current therapies for Parkinson's disease (PD) is the eventual manifestation of unpredictable motor fluctuations, ultimately resulting in alternating "on" periods with and without dyskinsesia and "off" periods (Deleu et al., 2002; Müller, 2002). PD patients, who show unpredictable or erratic on-off phenomena with oral levodopa therapy, have a predictable response to short-term intravenous administration (Baas et al., 1995). This suggests, that fluctuations of plasma concentrations of the given drug contribute to the onset and clinical manifestation of these levodopa associated long-term motor complications. Therefore a substained drug delivery, which may result in more continuous, constant plasma levels, may be a suitable therapeutic approach for improvement of these motor fluctuations in PD patients (Bennett Jr et al., 1989). Oral intake of compounds depends on gastrointestinal resorption with consecutive fluctuations of plasma concentrations. In contrast, transdermal drug delivery of dopamine substituting compounds enables more constant plasma levels, which may result in an amelioration of intensity and frequency of on-off phenomena (Metman et al., 2001; Behrens and Sommerville, 2003). However, the affinity of the administered dopaminergic compound to the postsynaptic dopaminergic receptor may also play an essential role for the onset of motor complications. In this respect, lisuride with its distinct higher affinity to postsynaptic dopaminergic striatal terminals compared with other dopamine agonists hypothetically represents an extraordinary suitable compound for the treatment of PD patients with motor complications (Rinne, 1999; Müller, 2002). Since lisuride has a relative short plasma halflife after oral intake, transdermal lisuride delivery may circumvent this disadvantage. Objective of this pilot study was to demonstrate the clinical efficacy, safety and tolerability of application of a novel transdermal lisuride formulation in PD patients with unpredictable motor fluctuations.

Patients and methods

Subjects

We enrolled 8 PD patients diagnosed according to UK Parkinson's disease Brain Bank criteria into this study (Hughes et al., 1992). Clinical characteristics are given in Table 1. All patients suffered from on-off fluctuations despite titration of the preexisting dopaminergic drug regime to its optimum efficacy and tolerability. All except one study participant finished the study after eight days. Subject MG withdraw her consent on day 4 due to administrative reasons.

Lisuride patches

The new developed employed patches (size = 15 cm^2) deliver 2–5 μg of lisuride base/cm^2/hour in mice. We attached each patch to the trunk for 48 hours.

Design

We increased the dosage on a day to day regimen. 3 patients (IO, HS, GK) received a dosage of up to 4 patches within the their treatment period, whereas the remaining 5 (GF, HM, MG, EB, CH) were titrated up to two patches (Table 1). All individuals were on a combination therapy consisting of various dopamine agonists and levodopa/DDI. We did not change the concomitant oral antiparkinsonian drug regimen during the study.

Assessment of clinical symptoms

We scored PD patients with the Hoehn and Yahr Scale (Hoehn and Yahr, 2001) and the Unified Parkinson's Disease Rating Scale (UPDRS) prior to study entry (Fahn et al., 1987). We used a self developed self assessment scale for scoring of motor function, which ranged from 1 (very good) to six (worst), similar to a patient diary (Poewe et al., 2002). We did not use the UPDRS part III in order to rate the severity of on- and off-periods for administrative reasons, since we simultaneously performed the self rating every 30 minutes.

Evaluation of the self assessment scale for motor function

We computed the motor changing rate (Δm), which we calculated according to the formula: Δm = change of motor function per interval/(number of scored intervals – 1). The ratings of patient IO illustrate our novel statistical approach (Figs. 1, 2). IO scored herself every 30 minutes with a total of 19 times. We calculated the change of motor function from one assessment timepoint to the next, therefore we computed 18 results. These values amounted to a total of 15 before lisuride patch application (Fig. 1). Correspondingly, we computed a Δm = 15/18 = 0.83. At the follow-up examination 8 days later, we computed a Δm = 4/18 = 0.22 (Fig. 2).

Table 1. Demographic and clinical characteristics of patients with Parkinson's disease

	Age	Sex	Weight	Height	PD duration	HYS	UPDRS	I	II	III	IV	On-off	Dyskinesia	Daily levodopa	Dopamine agonist (mg)
IO	61	F	51	163	10	3	63	2	16	41	4	X	X	375	lisuride (2.15)
HS	69	M	66	185	18	5	87	2	29	50	6	X	X	700	pergolide (1)
GK	63	F	58	165	12	4	61	4	24	27	6	X	X	325	pergolide (0.5); lisuride (0.4)
GF	61	M	76	174	12	2	30	1	4	21	4	X		1500	lisuride (1.5)
HM	68	M	84	189	14	4	79	0	20	52	7	X		725	lisuride (2)
MG	52	F	58	161	7	4	49	3	21	20	5	X	X	800	pramipexole (2.1)
EB	68	M	73	175	10	4	31	2	11	15	3	X	X	1300	pramipexole (2.8)
CH	67	F	46,7	167	32	3	46	1	23	19	3	X	X	625	pergolide (3)

Age is given in years; *PD duration* duration of Parkinson's disease in years; *HYS* Hoehn and Yahr Scale; *UPDRS* Unified Parkinson's Disease Rating Scale; *I* mental behaviour (items 1–4); *II* activities of daily living (items 5–17); *III* motor examination (items 18–31); *IV* complications of therapy (items 32–36); sex F(emale), M(ale), weight is given kg, height is given in cm, On-off = X (onset of on-, off-fluctuations), dyskinesia = X (onset); daily levodopa = daily levodopa dosage in mg, dopamine agonist (mg)

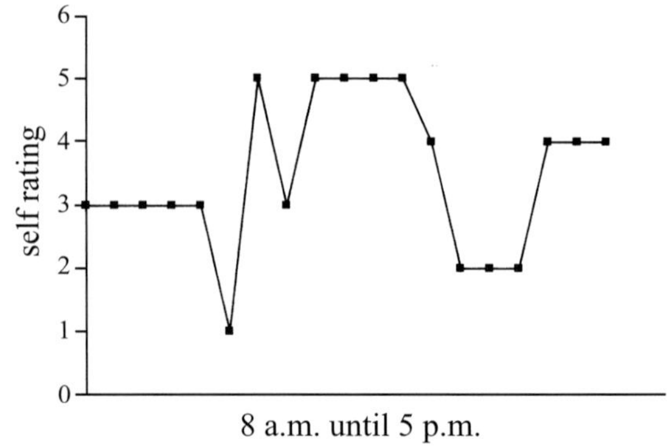

Fig. 1. Self assessment of IO on day 0 between 8 am and 5 pm

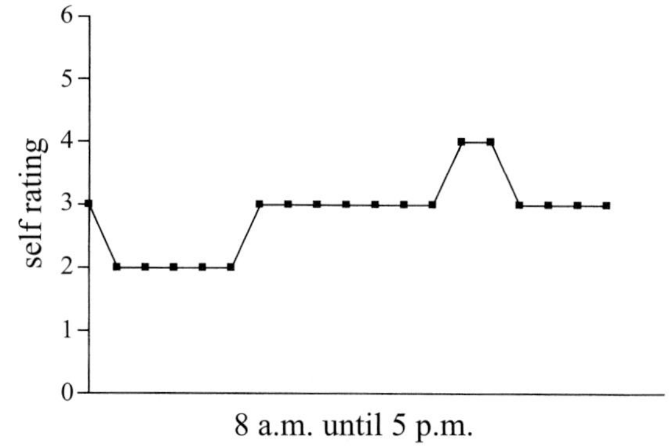

Fig. 2. Self assessment of IO on day 8 between 8 am and 5 pm, now treated with additional lisuride patches

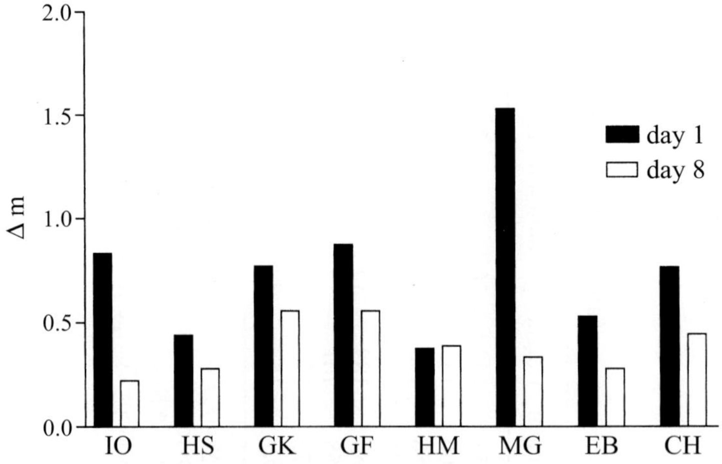

Fig. 3. Computed Δm on day 0 and day 8

Statistics

We compared the baseline data with the computed results on the last day of lisuride patch treatment (n = 7 on day 8, n = 1 on day 3). We used ANCOVA and set number of patches, age, sex, body weight, height, UPDRS total score as covariates. We used linear regression for

Table 2. Computed motor changing rate (Δm)

Day	IO	HS	GK	GF	HM	MG	EB	CH
0	0.83	0.44	0.77	0.88	0.38	1.53	0.53	0.76
1	0.39	0.73	0.50	0.56	0.72	0.72	1.11	0.50
2	0.60	0.50	0.78	0.39	0.50	0.83	0.67	0.22
3	0.72	0.28	0.29	0.72	0.39	0.33	0.78	0.19
4	0.50	0.22	0.44	0.67	0.28	ND	0.47	0.80
5	0.56	0.33	0.67	0.89	0.39	ND	ND	0.28
6	0.82	0.22	0.50	0.44	0.61	ND	0.54	0.83
7	0.67	0.39	0.33	0.72	0.39	ND	ND	0.50
8	0.22	0.28	0.56	0.56	0.39	ND	0.28	0.44

ND not done

correlation analysis between the computed difference before and after patch treatment and the various set covariates and adjusted the p-value according to the Bonferroni method to 0.007.

Ethics

All participants gave their written informed consent. The local ethics committee approved this trial.

Results

Δm

There was a significant improvement of the Δm (ANCOVA:$F_{(dF\ 1,\ dF\ 7)} = 8.45$, $p = 0.023$). Correlation analysis only showed significant inverse relations to the duration of transdermal lisuride delivery ($R = -0.88$, $p = 0.004$) and the age ($R = -0.91$, $p = 0.002$). There were no further significant associations between the computed differences and the employed covariates (results not shown). The individual Δm values of all patients are given in Table 2.

Adverse effects

There was a good tolerability of transdermal delivery of lisuride in all patients. Relevant side effects were local transient skin reactions at the attachment site in 3 participants. One female subject experienced skin irritations also beyond the patch application site (local red efflorescences and pustules) combined with edema, which disappeared despite ongoing treatment. The skin irritations dissappeared within 3 days after patch removal. There was no complain of nausea during the study.

Discussion

There has been a growing recognition that pulsatile stimulation of dopamine receptors may be an important mechanism in the generation of motor fluctuations that often develop and compromise the effectiveness of long-term levodopa administration in PD patients. Treatment approaches that might provide more constant, and thus physiological, dopamine receptor stimulation may reduce this phenomenon. Therefore frequent levodopa administration, controlled-release levodopa formulations, inhibitors of levodopa metabolism, duodenal, subcutaneous and even intravenous infusions of levodopa or dopamine agonists have all been employed, but all show limitations. In contrast, transdermal drug delivery is a treatment approach that is not only capable of providing a constant rate of drug delivery, but is also non-invasive and relatively simple to use (Pfeiffer, 2002).

The results of our study indicate that transdermal delivery of lisuride is safe, tolerable and efficacious, since it reduced the intensity and frequency of motor fluctuations in PD patients. Thus our preliminary results support the therapeutic concept of continuous dopaminergic stimulation for the treatment of on-off phenomena in patients with advanced PD (Calabrese et al., 1997; Montastruc et al., 1999; Metman et al., 2001; Hutton et al., 2001; Pfeiffer, 2002). An earlier trial on transdermal rotigotine application with a 2-week dose escalation phase followed by a 2-week dose maintenance phase at the highest dose showed similar results with regard to decrease of off periods, local skin reactions and symptomatic efficacy (Metman et al., 2001). Trials with transdermal delivery of levodopa, apomorphine and various other dopamine agonists are currently under investigation (Calabrese et al., 1997; Sudo et al., 1998; Danhof et al., 1998; Smith et al., 2000; Hutton et al., 2001). Similar to the other studies with transdermal delivery of antiparkinsonian drugs (Behrens et al., 2003), we found local skin reactions. The value and efficacy of our novel used simple statistical evaluation of motor fluctuations needs to be confirmed in future trials.

In conclusion our preliminary results indicate that transdermal lisuride delivery appears to be a promising future therapeutic concept for motor fluctuations in PD patients.

References

Baas H, Harder S, Demisch L, Burklin F, Stecker K, Fischer PA (1995) Fluctuations in Parkinson's disease. Pathogenetic significance of levodopa's cerebral pharmacokinetics and pharmacodynamics. J Neural Transm [Suppl] 46: 367–379

Behrens S, Sommerville K (2003) Non-oral drug delivery in Parkinson's disease: a summary from the symposium at the 7th International Congress of Parkinson's Disease and Movement Disorders, 10–14 November 2002, Miami, FL, USA. Expert Opin Pharmacother 4: 595–599

Bennett JP Jr, Turk M, Landow E (1989) Continuous oral administration of L-dihydroxyphenylalanine (L-DOPA) solution to patients with advanced Parkinson's disease. Clin Neuropharmacol 12: 285–292

Calabrese VP, Vaughan J, Lucas J, Hong AQ, Barrett R, Martin P, Sica D (1997) N-0923, a transdermal amino tetralin D2 dopamine agonist in the treatment of Parkinson's disease. Neurology 48: 32005

Danhof M, van der Geest R, van Laar T, Bodde HE (1998) An integrated pharmacokinetic-pharmacodynamic approach to optimization of R-apomorphine delivery in Parkinson's disease. Adv Drug Delivery Rev 33: 253–263

Deleu D, Northway MG, Hanssens Y (2002) Clinical pharmacokinetic and pharmacodynamic properties of drugs used in the treatment of Parkinson's disease. Clin Pharmacokinet 41: 261–309

Fahn S, Elton R, Members of the UPDRS Development Committee (1987) Unified Parkinson's Disease Rating Scale. Macmillan, New York, pp 153–163

Hoehn MM, Yahr MD (2001) Parkinsonism: onset, progression, and mortality. 1967. Neurology 57: S11–S26

Hughes AJ, Daniel SE, Kilford L, Lees AJ (1992) Accuracy of clinical diagnosis of idiopathic Parkinson's disease: a clinico-pathological study of 100 cases [see comments]. J Neurol Neurosurg Psychiatry 55: 181–184

Hutton JT, Metman LV, Chase TN, Juncos JL, Koller WC, Pahwa R, Lewitt PA, Samii A, Tsui JKC, Calne DB, Waters CH, Calabrese VP, Bennett JP, Barrett R, Morris JL (2001) Transdermal dopaminergic D-2 receptor agonist therapy in parkinson's disease with n-0923 TDS: a double-blind, placebo-controlled study. Mov Disord 16: 459–463

Metman LV, Gillespie M, Farmer C, Bibbiani F, Konitsiotis S, Morris M, Shill H, Bara-Jimenez W, Mouradian MM, Chase TN (2001) Continuous transdermal dopaminergic stimulation in advanced Parkinson's disease. Clin Neuropharmacol 24: 163–169

Montastruc JL, Ziegler M, Rascol O, Malbezin M (1999) A randomized, double-blind study of a skin patch of a dopaminergic agonist, piribedil, in Parkinson's disease. Mov Disord 14: 336–341

Müller T (2002) Dopaminergic substitution in Parkinson's disease. Expert Opin Pharmacother 3: 1393–1403

Pfeiffer RF (2002) Potential of transdermal drug delivery in Parkinson's disease. Drugs & Aging 19: 561–570

Poewe WH, Deuschl G, Gordin A, Kultalahti ER, Leinonen M (2002) Efficacy and safety of entacapone in Parkinson's disease patients with suboptimal levodopa response: a 6-month randomized placebo-controlled double-blind study in Germany and Austria (Celomen study). Acta Neurol Scand 105: 245–255

Rinne UK (1999) [Combination therapy with lisuride and L-dopa in the early stages of Parkinson's disease decreases and delays the development of motor fluctuations. Long-term study over 10 years in comparison with L-dopa monotherapy]. Nervenarzt 70 [Suppl 1]: S19–S25

Smith LA, Jackson MG, Bonhomme C, Chezaubernard C, Pearce RKB, Jenner P (2000) Transdermal administration of piribedil reverses MPTP-induced motor deficits in the common marmoset. Clin Neuropharmacol 23: 133–142

Sudo J, Iwase H, Terui J, Kakuno K, Soyama M, Takayama K, Nagai T (1998) Transdermal absorption of L-dopa from hydrogel in rats. Eur J Pharmaceut Sci 7: 67–71

Authors' address: D. Woitalla, MD, Department of Neurology, St. Josef Hospital, Ruhr University Bochum, Gudrunstrasse 56, D-44791 Bochum, Germany, e-mail: dirk.woitalla@ruhr-uni-bochum.de

J Neural Transm (2004) [Suppl] 68: 97–104

Congruent deterioration of complex and simple movements in patients with Huntington's disease

C. Saft, J. Andrich, N.-M. Meisel, H. Przuntek, and Th. Müller

Department of Neurology, St. Josef Hospital, Ruhr University, Bochum, Germany

Summary. Rating scales and assessment of simple and complex movements may reflect severity of Huntington's disease (HD). Objectives of our study were to compare scored HD symptoms and outcomes of instrumental tests, which demand for simple (tapping) and complex (peg insertion) movement series, in controls and subjects in various HD stages and to correlate them to each other. Motor test outcomes were significantly worsened in previously untreated and treated HD patients in comparison with HD gene carriers and controls. Peg insertion- and tapping results significantly correlated with the scored HD symptoms. Significant associations appeared between both motor test results in the controls, the previously untreated- and treated HD patients. Results of both instrumental tasks represent no specific diagnostic marker of HD, but the significant associations between both motor test outcomes indicate, that a parallel progress of deterioration of complex and simple movement abilities occurs after start of HD.

Introduction

A neurodegenerative, autosomal dominantly inherited disorder, which starts in early to middle adulthood, is Huntington's disease (HD). An expansion of a CAG repeat in the first exon of a gene on chromosome 4 encodes an expanded polyglutamine stretch in the N-terminal domain of a protein named Huntingtin, which accumulates and causes predominant apoptotic striatal neuronal death in HD (Furtado et al., 1996; de Boo et al., 1997; Sanchez-Pernaute et al., 2000; Paulsen et al., 2001). Clinical, neuropathologic, experimental studies support a major role of the CAG expansion in HD progression (Penney Jr et al., 1997; Sanchez-Pernaute et al., 2000). Characteristic clinical HD symptoms are psychiatric features, cognitive impairment and motor disturbances, all of which impair initiation, execution and reduction of movements (Penney Jr et al., 1997; Paulsen et al., 2001; Kirkwood et al., 2002). This altered movement behaviour causes a reduced performance of various valuable, at least partially rather complex instrumental motor tests (Rubin et al., 1993; Phillips et al., 1996; Thaut et al., 1999; van Vugt et al., 2001; Reilmann et al., 2001).

On the one hand these partial rather complex motor tests are not specific and sometimes not portable, but on the other hand this instrumental assessment of movement does not require the services of an experienced clinician (Folstein et al., 1983). A further rather simple method for the objective computer-based assessment of motor deficits is tapping (Müller et al., 2000, 2003). This test asks for repetitive performance and programming of standardized movements, whereas the more complex alternate finger tapping tasks (Pal et al., 2001) resemble to the standardized, easy-to-operate, computer based performance of peg insertion (Müller et al., 2000, 2003). Outcomes of peg insertion reflect the time period taken to insert 25 pegs from a rack into a series of appropriate holes. Thus this method avoids fatigue at the end of the test and circumvents the possibility that for some severe patients it would be impossible to complete the insertion of all 25 pegs in one column (Müller et al., 2000, 2003). This motor test resembles to the Purdue Pegboard, which counts the number of pegs inserted during a set period of 30 seconds for each hand (Müller et al., 2000). Both instrumental tools may aid assessment of motor performance, in particular bradykinesia, in neurodegenerative diseases, i.e. HD or Parkinson's disease (Müller et al., 2000; Garcia Ruiz et al., 2002) in addition to clinical rating with the Unified Huntington's Disease Rating Scale (UHDRS) and its subscores independence scale (IS), which represents a rough estimation of the HD subject's independence from the surroundings, and functional capacity (FC), which rates the HD subject's behaviour in daily life and the caregiver burden, to a certain extent (Huntington Study Group, 1996). Thus, this scale estimates behavioural, cognitive and motor dysfunction, but may suffer from the physician's subjective impression of the HD patient. This may influence the rating process and the score (Rubin et al., 1993; Phillips et al., 1996, 1996; Sanchez-Pernaute et al., 2000; Müller et al., 2000, 2003). Objectives of our study were to compare and to correlate outcomes of both motor tests to each other and their relation to the UHDRS scores of HD patients in various stages of the disease.

Subjects and methods

Subjects

We enrolled only right-handed, asymptomatic 22 HD gene carriers, 83 previously untreated HD patients with psychiatric or motor symptoms, or both, and 43 treated HD patients into this trial. All participants were consecutively referred to our department. In the group of treated HD patients, drug therapy consisted of tetrabenazine (N [number of subjects] = 17; 102.94 [mean] ± 73.76 [SD]; 25–350 [range] mg); tiapride (N = 32; 334.77 ± 243.52; 25–1200 mg); sulperide (N = 17; 188.24 ± 159.62; 50–600 mg); haloperidole (N = 1; 15 mg); memantine (N = 1; 20 mg); pipamperon (N = 1; 200 mg); risperidone (N = 1; 6 mg); perphenazine (N = 2; 4 mg & 16 mg) alone or in combination. Molecular analysis of the CAG expansion was performed, as previously described (Benitez et al., 1994). CAG significantly (ANOVA: $F_{(dF\,2,\,dF\,145)} = 4.71$, $p = 0.01$; post hoc analysis: HD gene carriers versus treated HD patients: $p = 0.02$; only significant results shown; Table 1) varied between the three groups of HD individuals (Table 1) The UHDRS total score significantly ($p < 0.0001$) differed between treated and untreated HD patients. We immediately scored HD patients before or after serial performance of both instrumental tasks with the UHDRS (Huntington Study Group, 1996). We blinded all participating physicians to each other. We also performed both instrumental tests in 57 right-handed healthy controls (Table 1).

Table 1. Clinical characteristics of HD subjects

	HD gene carrier	Untreated HD patients	Treated HD patients	Controls
Sex	16 women; 6 men	43 women; 40 men	21 women; 22 men	29 women; 28 men
Age	38.09 ± 8.12; 26–55	45.98 ± 8.77; 25–66	47.40 ± 10.44; 31–69	56.49 ± 11.32; 24–76
UHDRS		25.30 ± 12.02; 5–64	44.07 ± 14.35; 13–81	
FC		10.93 ± 1.98; 5–13	6.88 ± 2.25; 2–12	
IS		90.84 ± 9.9; 60–100	70.93 ± 12.5; 40–90	
CAG	42.95 ± 1.96; 40–47	45.12 ± 3.85; 38–63	45.84 ± 3.79; 40–57	
Tapping	354.41 ± 39.21; 290–452	273.28 ± 64.81; 113–397	201.56 ± 66.53; 89–410	360.44 ± 38.51; 250–454
Tapping right	186.18 ± 19.66; 149–223	152.64 ± 35.01; 61–231	113.72 ± 36.29; 54–216	187.91 ± 20.17; 131–227
Tapping left	168.23 ± 26.99; 127–229	120.64 ± 34.96; 39–201	87.84 ± 33.21; 31–194	172.53 ± 21.23; 119–231
Peg insertion	775.18 ± 54.08; 669–880	1195.06 ± 363.06; 708–3225	1698.63 ± 710.42; 1021–4231	838.28 ± 125.45; 633–1215
Peg insertion right	380.41 ± 37.84; 324–454	572.63 ± 181.62; 349–1515	782.23 ± 281.65; 494–1850	408.44 ± 70.13; 300–619
Peg insertion left	394.77 ± 26.47; 345–435	622.43 ± 192.86; 359–1710	916.40 ± 452.33; 521–2625	429.84 ± 60.93; 326–596

CAG CAG repeat length; *FC* Score of the Functional Capacity of the UHDRS; *IS* Score of the Independence Scale of the UHDRS; HD duration is given in years; *UHDRS* Unified Huntington's Disease Rating Scale (UHDRS) score; time intervals for peg insertion are given in milliseconds; peg insertion (tapping) (right/left) = peg insertion (tapping) outcome with the right/left hand; peg insertion (tapping) = computed sum of both hands; all data are given as mean ± standard deviation, range

Peg insertion

We asked subjects to transfer 25 pegs (diameter 2.5 mm, length 5 cm) from a rack into one of 25 holes (diameter 2.8 mm) in a computer-based contact board individually and as quickly as possible. The distance between rack and appropriate holes was 32 cm. The board was positioned in the middle and the task was carried out on each side. When transferring each peg from rack to hole, elbows were allowed to be in contact with the table. We measured the time interval between inserting of the first and the last pin initially with the right- and then the left hand. We assessed the time period for this task by a computer to 100 milliseconds (ms) accuracy. The total peg insertion score is the sum of the time intervals of task performance with the right and left hand in ms (Table 1) (Müller et al., 2000).

Tapping

Individuals as quickly as possible tapped on a contact board (3 cm × 3 cm) with a contact pencil for a period of 32 seconds after the initial flash of a yellow stimulus light. The board was in a centralized position, when the task was carried on each side. When performing the tapping, elbows were allowed to be in contact with the table. We obtained the number of contacts by a computer. First we measured the frequency of tapping with the right and then with the left hand. Tapping represents the sum of tapping results of the right and left hand (Table 1).

We intended to minimize learning and training effects. Therefore we allowed all subjects to familiarize with this test for one minute on the day before the start of the study.

Statistics

Data showed a normal distribution according to the Kolmogorow-Smirnow test. Consequently, we performed parametric tests. We used ANCOVA with sex and age as covariates for comparisons in combination with the Tukeys HSD-test for different numbers for the post hoc analysis. We employed Pearson Product-Moment correlations for correlation analysis.

Ethics

We obtained informed consent from each subject. The local ethics committee of the university approved this study.

Results

Comparisons

Intervals for the peg insertion task significantly (both hands: ANCOVA: $F_{(dF\,3,\,dF\,199)} = 43.89$, $p < 0.0001$; right: $F_{(dF\,3,\,dF\,199)} = 44.74$, $p < 0.0001$; left: $F_{(dF\,3,\,dF\,199)} = 38.23$, $p < 0.0001$) differed between previously untreated and treated HD patients, HD gene carriers and controls (Table 1, post hoc analysis see Table 2). An impact of the covariates did not appear.

Accordingly, significantly (both hands: ANCOVA: $F_{(dF\,3,\,dF\,199)} = 83.27$, $p < 0.0001$; right: $F_{(dF\,3,\,dF\,199)} = 60.60$, $p < 0.0001$; left: $F_{(dF\,3,\,dF\,199)} = 83.97$, $p < 0.0001$) varying tapping rates appeared between the various investigated groups of individuals (Table 1, post hoc analysis see Table 2). There was a significant impact of the covariates sex and age (Age: β-coefficient [β-c] = −0.85, p = 0.03; sex: β-c = −23.49, p = 0.003) on the tapping results.

Correlation analysis

There were significant relations between the peg insertion intervals and tapping rates in the controls, the previously untreated- and treated HD patients, but not in the HD gene carriers (Table 3).

Table 2. Post hoc analysis of both instrumental tests

Comparison	Peg insertion			Tapping		
	both hands	right land	left hand	both hands	right land	left hand
Controls vs HD gene carrier	0.95	0.95	0.96	0.98	0.99	0.96
Controls vs untreated HD patients	**0.003**	**0.002**	**0.01**	**1.38E−05**	**0.001**	**8.29E−06**
Controls vs treated HD patients	**7.69E−06**	**7.69E−06**	**7.69E−06**	**7.69E−06**	**7.69E−06**	**7.69E−06**

p-values of the Tukeys HSD test for different numbers, significant comparisons are bold, *vs* versus

Table 3. Correlation analysis of instrumental test outcomes

Correlation of tapping- and peg insertion results		
	R	p
HD gene carriers		
both hands	−0.24	0.28
right hand	−0.33	0.14
left hand	0.06	0.77
Untreated HD patients		
both hands	**−0.66**	**5.11E−11**
right hand	**−0.60**	**2.18E−09**
left hand	**−0.60**	**1.41E−09**
Treated HD patients		
both hands	**−0.59**	**3.25E−05**
right hand	**−0.59**	**3.04E−05**
left hand	**−0.52**	**0.0003**
Controls		
both hands	**−0.51**	**5E−05**
right hand	**−0.52**	**2.8E−05**
left hand	**−0.42**	**0.001**

p p-values of the correlation analysis; *R* correlation coefficient, significant results are bold

The computed sum of peg insertion outcomes and tapping results, performed with the right and left hand separately, significantly correlated with the UHDRS score in the treated (peg insertion: $R = 0.48$; $p = 0.001$; tapping: $R = -0.43$; $p = 0.004$) and previously untreated (peg insertion: $R = 0.65$; $p < 0.0001$; tapping: $R = -0.66$; $p < 0.0001$) HD patients.

There was a significant association between the analogously computed peg insertion intervals, respectively tapping rates, and the FC score in the previously untreated- (peg insertion: $R = -0.52$; $p < 0.0001$; tapping: $R = 0.43$; $p < 0.0001$) and treated (peg insertion: $R = -0.55$; $p = 0.0001$; tapping: $R = 0.37$; $p = 0.015$) HD individuals, but not in the HD gene carriers (results not shown).

Analogously, we also obtained significant associations between motor test outcomes and the IS score in the previously untreated- (peg insertion: $R = -0.46$; $p < 0.0001$; tapping: $R = 0.42$; $p < 0.0001$) and treated (peg insertion: $R = -0.63$; $p < 0.0001$; tapping: $R = 0.34$; $p = 0.02$) HD individuals, but not in the HD gene carriers (results not shown).

Correlation analysis between dosages of applied compounds and results of both motor tests, if applicable, showed no significant outcomes (results not shown).

Discussion

There was a certain overlap of motor test outcomes between the various types of HD individuals. Nevertheless the performed standardized measurements of complex movement series with the peg insertion task and simple motion sequences with the tapping test reflected abnormalities of motor behaviour, respectively scored HD symptoms, in HD patients in contrast to the HD gene carriers and controls despite the limitation of both motor test performance to the upper extremities. Results of both tests correlate to each other in HD patients and controls, although they hypothetically demand for different levels and efforts of cognitive processes. The peg insertion procedure needs more cognitive load due to the concomitant additional involvement of visuospatial cognition, self-elaboration of internal strategies, sorting and planning. All these processes are influenced by the modulatory role of striatal dopamine levels on association areas of the prefrontal cortex (Lalonde and Botez-Marquard, 1997; Nieoullon, 2002; Müller et al., 2003). Thus striatal dopaminergic neurotransmission does not only play a key role in movement behaviour, it also provides a certain indirect impact on cognitive function and regulates attention via the frontal lobe and basal ganglia (Lalonde et al., 1997; Lawrence et al., 1998; Sanchez-Pernaute et al., 2000; Nieoullon, 2002). Thus dopamine acts as a powerful regulator of different aspects of cognitive brain functions, which has also been demonstrated by the onset of cognitive deficits after selective lesioning of mesocorticolimbic components of the dopaminergic system (Nieoullon, 2002). We assume, that performance of peg insertion involves all these subcortical regions. In view of our results, we suggest, that the peg insertion task also reflects this altered executive dysfunction with impaired planning and performance of complex movements due to fronto-striatal pathology in early HD (Paulsen et al., 2001; Redondo et al., 2001; Kirkwood et al., 2002; Müller et al., 2003). In contrast, the tapping paradigm asks for low cognitive efforts, since the subject may create a fixed habit tendency with a consistent, attentional saving of cognitive resources due to the repetitive performance and programming of a standardized movement sequence based on an automatic function of a cognitive set (Brown and Marsden, 1990; Lalonde et al., 1997). Thus the tapping procedure asks for the ability of autonomic functioning of the attentional control (Müller et al., 2003). This selective attentional processing reduces during HD progression (Claus and Mohr, 1996). Therefore tapping rates may relate to this increasing cognitive impairment in HD patients (Claus et al., 1996). A PET study with repeated measurements of regional cerebral blood flow consistently located the functional neuroanatomy of similar tapping movements to the precentral and postcentral gyri and supplementary motor area in younger subjects (Carey et al., 2000). Our results hypothetically indicate, that the deterioration of tapping numbers may partially also reflect the neurodegenerative process particularly of the primary sensorimotor cortex and supplementary motor area in HD (Carey et al., 2000).

Since a correlation between peg insertion- and tapping outcomes did not occur in the HD gene carriers, we assume, that this may reflect a certain dysbalance of the congruent regulation of simple and complex motion sequences, as shown in healthy controls in our trial. Therefore we suggest, that this missing association may hypothetically represent an initial sign of HD onset far beyond the recognition of clinical symptoms, although both, results of peg insertion and tapping, did not significantly differ to the results of the controls.

The distinctly impaired instrumental test outcomes of the treated more advanced HD patients result from the clinical motor symptoms, but they may also additionally be biased by the chronic drug regimen, which also included neuroleptics with their marked impact on movement. A further trial is necessary to find out the degree of this putative drug effect on the motor task results. We found no significant outcomes in our analysis, but we did not consider the acute effects of drug intake on motor test performance. Moreover some HD patients were on a combination drug regime, which does not allow to determine the exact effect of each compound on the instrumental test result. An impact of age on fine motor behaviour is known in healthy individuals (Müller et al., 2000), but we assume, that the individual different disturbed fine motor control hypothetically caused the missing age dependence in HD patients. In the HD gene carriers, the missing significant relation between age and test outcomes could hypothetically result from an evolving reduced capability of movement performance (results not shown). The significant relations between motor test results and the IS- and the FC-subscores in the HD patients indicate, that both instrumental tools are also able to reflect the HD subject's independence in daily life and the caregiver burden due to HD.

In conclusion, the results of both instrumental tasks represent no specific diagnostic marker for HD, since other diseases and/or fatigue may also influence movement behaviour and motor control. But the significant associations between both motor test outcomes indicate, that there is a certain parallel progress of deterioration of complex and simple movement behaviour abilities after onset of HD. Future studies are necessary to provide further information on the ability and quality of both tests to follow intra patient responses to therapeutic interventions or HD progression.

References

Anonymous (1996) Unified Huntington's Disease Rating Scale: reliability and consistency. Huntington Study Group. Mov Disord 11: 136–142

Benitez J, Fernandez E, Garcia RP, Robledo M, Ramos C, Yebenes J (1994) Trinucleotide (CAG) repeat expansion in chromosomes of Spanish patients with Huntington's disease. Hum Genet 94: 563–564

Brown RG, Marsden CD (1990) Cognitive function in Parkinson's disease: from description to theory. Trends Neurosci 13: 21–29

Carey LM, Abbott DF, Egan GF, Tochon-Danguy HJ, Donnan GA (2000) The functional neuroanatomy and long-term reproducibility of brain activation associated with a simple finger tapping task in older healthy volunteers: a serial PET study. Neuroimage 11: 124–144

Claus JJ, Mohr E (1996) Attentional deficits in Alzheimer's, Parkinson's, and Huntington's diseases. Acta Neurol Scand 93: 346–351

de Boo GM, Tibben A, Lanser JB, Jennekens-Schinkel A, Hermans J, Maat-Kievit A, Roos RA (1997) Early cognitive and motor symptoms in identified carriers of the gene for Huntington disease. Arch Neurol 54: 1353–1357

Folstein SE, Jensen B, Leigh RJ, Folstein MF (1983) The measurement of abnormal movement: methods developed for Huntington's disease. Neurobehav Toxicol Teratol 5: 605–609

Furtado S, Suchowersky O, Rewcastle B, Graham L, Klimek ML, Garber A (1996) Relationship between trinucleotide repeats and neuropathological changes in Huntington's disease. Ann Neurol 39: 132–136

Garcia Ruiz PJ, Hernandez J, Cantarero S, Bartolome M, Sanchez BV, Garcia dY (2002) Bradykinesia in Huntington's disease. A prospective, follow-up study. J Neurol 249: 437–440

Kirkwood SC, Siemers E, Viken R, Hodes ME, Conneally PM, Christian JC, Foroud T (2002) Longitudinal personality changes among presymptomatic Huntington disease gene carriers. Neuropsychiatry Neuropsychol Behav Neurol 15: 192–197

Lalonde R, Botez-Marquard T (1997) The neurobiological basis of movement initiation. Rev Neurosci 8: 35–54

Lawrence AD, Weeks RA, Brooks DJ, Andrews TC, Watkins LH, Harding AE, Robbins TW, Sahakian BJ (1998) The relationship between striatal dopamine receptor binding and cognitive performance in Huntington's disease. Brain 121 (Pt 7): 1343–1355

Müller T, Schäfer S, Kuhn W, Przuntek H (2000) Correlation between tapping and inserting of pegs in Parkinson's disease. Can J Neurol Sci 27: 311–315

Müller T, Kuhn W, Schulte T, Przuntek H (2003) Intravenous amantadine sulphate application improves the performance of complex but not simple motor tasks in patients with Parkinson's disease. Neurosci Lett 339: 25–28

Nieoullon A (2002) Dopamine and the regulation of cognition and attention. Prog Neurobiol 67: 53–83

Pal PK, Lee CS, Samii A, Schulzer M, Stoessl AJ, Mak EK, Wudel J, Dobko T, Tsui JK (2001) Alternating two finger tapping with contralateral activation is an objective measure of clinical severity in Parkinson's disease and correlates with PET. Parkinsonism Rel Disord 7: 305–309

Paulsen JS, Zhao H, Stout JC, Brinkman RR, Guttman M, Ross CA, Como P, Manning C, Hayden MR, Shoulson I (2001) Clinical markers of early disease in persons near onset of Huntington's disease. Neurology 57: 658–662

Penney JB Jr, Vonsattel JP, MacDonald ME, Gusella JF, Myers RH (1997) CAG repeat number governs the development rate of pathology in Huntington's disease. Ann Neurol 41: 689–692

Phillips JG, Bradshaw JL, Chiu E, Teasdale N, Iansek R, Bradshaw JA (1996) Bradykinesia and movement precision in Huntington's disease. Neuropsychologia 34: 1241–1245

Redondo VL, Brown RG, Chacon P Jr (2001) [Executive dysfunction in Huntington's disease]. Rev Neurol 32: 923–929

Reilmann R, Kirsten F, Quinn L, Henningsen H, Marder K, Gordon AM (2001) Objective assessment of progression in Huntington's disease: a 3-year follow-up study. Neurology 57: 920–924

Rubin AJ, King WM, Reinbold KA, Shoulson I (1993) Quantitative longitudinal assessment of saccades in Huntington's disease. J Clin Neuroophthalmol 13: 59–66

Sanchez-Pernaute R, Kunig G, del Barrio AA, de Yebenes JG, Vontobel P, Leenders KL (2000) Bradykinesia in early Huntington's disease. Neurology 54: 119–125

Thaut MH, Miltner R, Lange HW, Hurt CP, Hoemberg V (1999) Velocity modulation and rhythmic synchronization of gait in Huntington's disease. Mov Disord 14: 808–819

van Vugt JP, Siesling S, Piet KK, Zwinderman AH, Middelkoop HA, van Hilten JJ, Roos RA (2001) Quantitative assessment of daytime motor activity provides a responsive measure of functional decline in patients with Huntington's disease. Mov Disord 16: 481–488

Authors' address: Th. Müller, MD, Department of Neurology, St. Josef Hospital, Ruhr University Bochum, Gudrunstrasse 56, D-44791 Bochum, e-mail: thomas.mueller@ruhr-uni-bochum.de

J Neural Transm (2004) [Suppl] 68: 105–110

Methyltetrahydrofolate reductase polymorphism influences onset of Huntington's disease

N. Brune[1], **J. Andrich**[1], **M. Gencik**[2], **C. Saft**[1], **Th. Müller**[1], **S. Valentin**[2], **H. Przuntek**[1], and **J. T. Epplen**[2]

[1] Department of Neurology, St. Josef Hospital, and
[2] Department of Human Genetics, Ruhr University, Bochum, Germany

Summary. Onset of Huntington's disease (HD) negatively correlates with CAG repeat length of the HD gene, which encodes the protein huntingtin. This protein interacts with the homocysteine metabolizing enzyme cystathionine β-synthase (CBS). Objective of this study was to analyze the impact of CAG repeats, polymorphisms of various homocysteine metabolizing enzymes, like CBS, Methyltetrahydrofolate Reductase (MTHTR), Methionine Synthase Reductase (MSR) and methionine synthase (MS) on HD onset in 171 patients. The significant impact of CAG repeats on HD onset ($\chi^2 = 25.54$, FG = 4, $p < 0.0001$) with a significant correlation between both ($R = -0.521$, $p = 0.01$) was obvious. HD patients with the homozygous MTHFR-1298-CC significantly ($p = 0.024$) earlier experienced HD symptoms. There was no influence demonstrable of CBS, MSR and MS. Determination of MTHFR polymorphisms and CAG repeats enables screening for subjects with putative early HD onset in order to study neuroprotective compounds in their efficacy to delay HD symptoms.

Introduction

Huntington's disease (HD) is a neurodegenerative, autosomal dominantly inherited disorder, which manifests usually in early to middle adulthood. HD is associated with an expansion of a CAG repeat in the first exon of a gene on chromosome 4. Available information from clinical, neuropathologic, and experimental studies support a major role of the CAG expansion in the onset and progression of disease (Penney et al., 1997; Sanchez-Pernaute et al., 2000). The mutant HD gene encodes an expanded polyglutamine stretch in the N-terminal domain of a protein named huntingtin, which accumulates and causes apoptotic striatal neuronal death (Furtado et al., 1996; de Boo et al., 1997; Sanchez-Pernaute et al., 2000; Paulsen et al., 2001). However, a further neurodegeneration promoting mechanism in HD is increased excitotoxicity, which may be at least partially triggered by huntingtin fragments in the striatum occuring elevated inclusion bodies in post mortem HD brain tissue (May and Gray, 1985; Schilling et al., 1999). Huntingtin interacts with the homocystinuria inducing enzyme cystathionine beta-synthase (CBS) (Boutell et al., 1998). This may contribute to total homocysteine elevation in plasma of HD patients (Andrich et al., 2004). Homocysteine shows glutamate like excitotoxic effects in neuronal cell culture models (May et al., 1985; Kruman et al., 2000). Thus homocysteine may hypothetically enhance glutamatergic cortical projection neurons, which finally results in Caspase-3 activation (Kruman et al., 2000). This triggers a sequential proteolysis of huntingtin with consecutive accumulation of N-terminal mutant fragments in the neuronal nuclei. Involved regulatory enzymes of homocysteine metabolism are methyltetrahydrofolate reductase (MTHFR) (Goyette et al., 1998, OMIM:*607093), CBS (Kraus et al., 1998, OMIM:*236200), methionine synthase reductase (MSR) (Leclerc et al., 1995, OMIM:*602568) and methionine synthase (MS) (Watkins et al., 2002, OMIM:*156570). This suggests, that mutations of MTHFR, CBS and MS may influence age of HD onset, similarly to the known impact of CAG repeat length. Objective of this study was to analyze the impact of CAG repeat length and enzyme polymorphisms of homocysteine metabolism on age of HD onset.

Subjects and methods

Table 1 shows some characteristics of the enrolled HD patients with a CAG repeat length between 41–45 (Mean: 43 ± 1.27 SD). We used polymerase chain reaction based technology and restriction enzymes for analysis of CAG repeat length, MTHFR, CBS, MSR and MS polymorphisms (Boutell et al., 1998; Geisel et al., 2001).

Statistics

We employed the Kruskal Wallis ANOVA, respectively the Mann Whitney U test for comparisons and Spearman rank correlation for correlation analysis.

Ethics

All participants gave written informed consent. The local ethics committee of the University approved the study.

Table 1. Selected clinical characteristics of patients with Huntington's disease

	Age of onset	Sex
CAG 41	52.7 (9.9)	11 men; 12 women
CAG 42	48.7 (10.3)	18 men; 23 women
CAG 43	44.6 (8.0)	26 men, 23 women
CAG 44	43.9 (7.9)	13 men; 18 women
CAG 45	38.4 (6.3)	14 men; 13 women

Data of age of onset are given in mean (standard deviation) in years, *CAG n* number of CAG repeats

Table 2. Polymorphisms of homocysteine metabolizing enzymes and age of HD onset

N	Genotype	Age of onset	CAG	p
29	MTHFR-677-CC (WT)	44.8 (10.3)	43.0 (1.2)	n.s.
71	MTHFR-677-CT	44.9 (10.1)	42.8 (1.1)	n.s.
21	MTHFR-677-TT	45.5 (7.2)	43.3 (1.4)	n.s.
75	MTHFR-1298-AA (WT)	46.0 (9.0)	42.9 (1.2)	n.s.
72	MTHFR-1298-AC	45.9 (8.8)	43.1 (1.2)	n.s.
11	MTHFR-1298-CC	41.3 (12.1)	43.1 (1.5)	0.024
138	CBS-68bp −/−	47.3 (8.6)	42.9 (1.3)	n.s.
22	CBS-68bp +/−	44.0 (11.8)	43.1 (1.2)	n.s.
61	MSR-66-AA (WT)	44.6 (9.1)	43.1 (1.2)	n.s.
76	MSR-66-AG	45.8 (10.1)	42.8 (1.3)	n.s.
23	MSR-66-GG	46.7 (8.3)	43.0 (1.1)	n.s.
102	MS-2756-AA (WT)	45.5 (9.8)	43.1 (1.2)	n.s.
54	MS-2756-AG	45.4 (8.7)	42.8 (1.2)	n.s.
4	MS-2756-GG	46.2 (14.5)	43.0 (1.9)	n.s.

N number of HD patients; *MTHFR* methyltetrahydrofolate reductase; *CBS* cystathionine β-synthase; *MSR* methionine synthase reductase; *MS* methionine synthase; *WT* wild type; *CAG* number of CAG repeats given in mean (standard deviation); data of age of onset are given in mean (standard deviation) in years, p = p-value of the comparisons performed with the Mann Whitney U test

Results

There was a significant impact of CAG repeat length on age of HD onset ($\chi^2 = 25.54$, FG = 4, $p < 0.0001$ [significant ($p < 0.0001$) post hoc comparisons: age of HD onset with CAG 41 versus CAG 43, 44, 45; age of HD onset with CAG 42 versus 43, 44, 45; age of HD onset with CAG 43 versus 45, age of HD onset with CAG 44 versus 45]) (Table 1). Accordingly, there was a significant correlation between both ($R = -0.521$, $p = 0.01$).

Table 2 shows the results of the analysis for the various genetic subtypes of the enzymes MTHFR, CBS, MSR and MS and the corresponding age of HD onset. HD patients with the MTHFR-1298-CC constitution showed a distinctly lower age of onset compared to the wild type genotypes.

Discussion

Our study confirms the negative correlation of age of HD onset and CAG repeat length (Benitez et al., 1994; Penney et al., 1997). We found no influence of genetic differences in the MSR, MS and CBS enzymes, on the appearance of HD symptoms. However, HD subjects with a homozygous MTHFR-1298-CC genotype experienced start of HD significantly earlier. There was no bias of CAG distribution on this result, since we found no significant (results not shown, Table 2) different CAG variance in the analysed subgroups of the MTHFR-1298-CC polymorphism. There is a need to confirm our preliminary findings in a larger cohort of HD patients due to the low number of subjects carrying the MTHFR-1298-CC variant. Nevertheless we suggest, that genetic

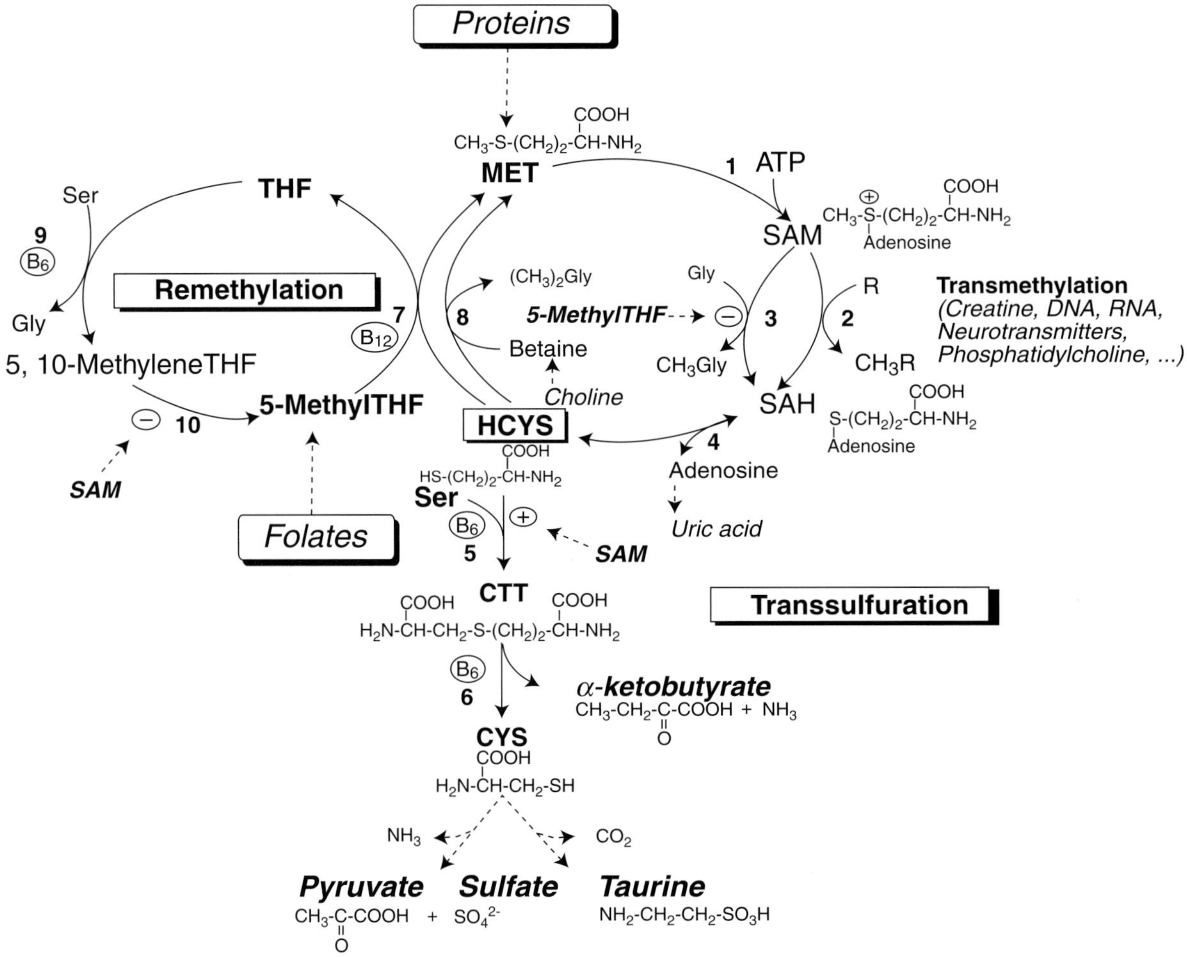

Fig. 1. Homocysteine metabolism. (**1**) Activation of methionine (*MET*) by ATP in the presence of methionine S-adenosyl-transferase. (**2**) S-adenosylmethionine (*SAM*) demethylation into S-adenosylhomocysteine (*SAH*) coupled with methylation of an acceptor R into RCH_3. (**3**) Alternative demethylation catalyzed by glycine N-methyltransferase, which converts glycine (*Gly*) into sarcosine (CH_3Gly). (**4**) Hydrolysis of SAH into homocysteine (*HCYS*) and adenosine catalyzed by S-adenosylhomocysteine hydrolase. (**5**) Condensation of serine (*Ser*) with HCYS to form cystathionine (*CTT*) through the action of vitamin B_6-derived pyridoxal-5′-phosphate-dependent **cystathionine ß-synthase** (***CBS***). (**6**) Conversion of CTT into cysteine (*CYS*) and α-ketobutyrate catalyzed by pyridoxal-5′-phosphate-dependent γ-cystathionase. (**7**) N^5-methyltetrahydrofolate (5-methylTHF) demethylation into THF and HCYS remethylation into MET by **methionine synthase** (**MS**) dependent on vitamin B_{12}-derived methylcobalamin. Within this reaction the MS is oxidated, so **methionine synthase reductase** (**MSR**) reduces this essential enzyme. (**8**) In the liver, remethylation of HCYS by betaine-homocysteine methyltransferase in the presence of choline-derived betaine. (**9**) Synthesis of $N^{5,10}$-methylenetetrahydrofolate (5, 10-methyleneTHF) from THF coupled with the conversion of Ser into Gly through the action of pyridoxal-5′-phosphate-dependent serine hydroxymethyltransferase. (**10**) Reduction of 5,10-methyleneTHF into 5-methylTHF catalyzed by **$N^{5,10}$-methylenetetrahydrofolate reductase** (***MTHFR***)(adapted from Durand et al., 2001)

testing for the MTHFR polymorphism in addition to the CAG repeat length would make sense in order to study putative neuroprotective compounds in their efficacy to delay onset of HD symptoms. The MTHFR-1298-CC polymorphism is associated with reduced enzyme activity between 50 and 60% in homozygous individuals and consequently it increases total homocysteine plasma levels (Geisel et al., 2001). Homocysteine may induce endothelial dysfunction and may further promote susceptibility to impaired mitochondrial energy metabolism beside the known glutamate mediated excitotoxicity (Chambers et al., 1999). Our results indicate, that particularly HD patients with the MTHFR-1298-CC genotype possibly benefit from long-term coenzyme Q10 and remacemide application, both of which showed neuroprotective effects in transgenic mouse models of HD (Ferrante et al., 2002). Moreover, we suggest monitoring of total plasma homocysteine concentrations and folic acid supplements for concomitant homocysteine lowering in these subtype of HD patients, since folic acid catalyzes and enhances metabolism of homocysteine to methionine (Malinow et al., 1998; Chambers et al., 1999; Malinow, 1999). Methionine acts in combination with pyridoxalphosphate or S-methyl-α-keto-butyric acid as strong scavenger of oxidants, which in turn induce endothelial dysfunction and atherosclerosis related disorders in the long-term (Malinow et al., 1998; Malinow, 1999). We suggest a future long-term trial on additional folic acid supplementation in HD patients with the homozygous MTHFR-1298-CC mutation in order to demonstrate delayed start of HD and reduction of elevated epidemiologic incidence of ischaemic heart and cerebrovascular disease beside pneumonia in HD patients (Lanska et al., 1988).

In conclusion our study confirms again, the decisive impact of CAG repeats on appearance of HD symptoms. In addition, we submit that the excitotoxic homocysteine metabolizing enzyme MTHFR influences the age of onset in HD.

References

Andrich J, Saft C, Arz A, Schneider B, Agelink MW, Kraus PH, Kuhn W, Müller T (2004) Hyperhomocysteinaemia in treated patients with Huntington's disease. Mov Disord 19: 226–228

Benitez J, Fernandez E, Garcia RP, Robledo M, Ramos C, Yebenes J (1994) Trinucleotide (CAG) repeat expansion in chromosomes of Spanish patients with Huntington's disease. Hum Genet 94: 563–564

Boutell JM, Wood JD, Harper PS, Jones AL (1998) Huntingtin interacts with cystathionine beta-synthase. Hum Mol Genet 7: 371–378

Chambers JC, McGregor A, Jean-Marie J, Kooner JS (1999) Abnormalities of vascular endothelial function may contribute to increased coronary heart disease risk in UK Indian Asians. Heart 81: 501–504

Durand P, Prost M, Loreau N, Lussier-Cancan S, Blache D (2001) Impaired homocysteine metabolism and atherothrombotic disease. Lab Invest 81: 645–672

de Boo GM, Tibben A, Lanser JB, Jennekens-Schinkel A, Hermans J, Maat-Kievit A, Roos RA (1997) Early cognitive and motor symptoms in identified carriers of the gene for Huntington disease. Arch Neurol 54: 1353–1357

Ferrante RJ, Andreassen OA, Dedeoglu A, Ferrante KL, Jenkins BG, Hersch SM, Beal MF (2002) Therapeutic effects of coenzyme Q10 and remacemide in transgenic mouse models of Huntington's disease. J Neurosci 22: 1592–1599

Furtado S, Suchowersky O, Rewcastle B, Graham L, Klimek ML, Garber A (1996) Relationship between trinucleotide repeats and neuropathological changes in Huntington's disease. Ann Neurol 39: 132–136

Goyette P, Pai A, Milos R, Frosst P, Tran P, Chen Z, Chan M, Rozen R (1998) Gene structure of human and mouse methylenetetrahydrofolate reductase (MTHFR). Mammalian Genome 9: 652–656

Geisel J, Zimbelmann I, Schorr H, Knapp JP, Bodis M, Hubner U, Herrmann W (2001) Genetic defects as important factors for moderate hyperhomocysteinemia. Clin Chem Lab Med 39: 698–704

Kraus JP, Oliveriusova J, Sokolova J, Kraus E, Vlcek C, de Franchis R, Maclean KN, Bao L, Bukovska G, Patterson D, Paces V, Ansorge W, Kozich V (1998) The human cystathionine beta-synthase (CBS) gene: complete sequence, alternative splicing, and polymorphisms. Genomics 52: 312–324

Kruman II, Culmsee C, Chan SL, Kruman Y, Guo Z, Penix L, Mattson MP (2000) Homocysteine elicits a DNA damage response in neurons that promotes apoptosis and hypersensitivity to excitotoxicity. J Neurosci 20: 6920–6926

Lanska DJ, Lavine L, Lanska MJ, Schoenberg BS (1988) Huntington's disease mortality in the United States. Neurology 38: 769–772

Leclerc D, Wilson A, Dumas R, Gafuik C, Song D, Watkins D, Heng HHQ, Rommens JM, Scherer SW, Rosenblatt DS, Gravel RA (1998) Cloning and mapping of a cDNA for methionine synthase reductase, a flavoprotein defective in patients with homocystinuria. Proc Natl Acad Sci 95: 3059–3064

Malinow MR (1999) Homocyst(e)ine, vitamins and genetic interactions in vascular disease. Can J Cardiol 15 [Suppl B]: 31B–34B

Malinow MR, Duell PB, Hess DL, Anderson PH, Kruger WD, Phillipson BE, Gluckman RA, Block PC, Upson BM (1998) Reduction of plasma homocyst(e)ine levels by breakfast cereal fortified with folic acid in patients with coronary heart disease [see comments]. N Engl J Med 338: 1009–1015

May PC, Gray PN (1985) L-Homocysteic acid as an alternative cytotoxin for studying glutamate-induced cellular degeneration of Huntington's disease and normal skin fibroblasts. Life Sci 37: 1483–1489

Paulsen JS, Zhao H, Stout JC, Brinkman RR, Guttman M, Ross CA, Como P, Manning C, Hayden MR, Shoulson I (2001) Clinical markers of early disease in persons near onset of Huntington's disease. Neurology 57: 658–662

Penney Jr JB, Vonsattel JP, MacDonald ME, Gusella JF, Myers RH (1997) CAG repeat number governs the development rate of pathology in Huntington's disease. Ann Neurol 41: 689–692

Sanchez-Pernaute R, Kunig G, del Barrio AA, de Yebenes JG, Vontobel P, Leenders KL (2000) Bradykinesia in early Huntington's disease. Neurology 54: 119–125

Schilling G, Becher MW, Sharp AH, Jinnah HA, Duan K, Kotzuk JA, Slunt HH, Ratovitski T, Cooper JK, Jenkins NA, Copeland NG, Price DL, Ross CA, Borchelt DR (1999) Intranuclear inclusions and neuritic aggregates in transgenic mice expressing a mutant N-terminal fragment of huntingtin. Hum Mol Genet 8: 397–407

Watkins D, Ru M, Hwang HY, Kim CD, Murray A, Philip NS, Kim W, Legakis H, Wai T, Hilton JF, Ge B, Dore C, Hosack A, Wilson A, Gravel RA, Shane B, Hudson TJ, Rosenblatt DS (2002) Hyperhomocysteinemia due to methionine synthase deficiency, cblG: structure of the MTR gene, genotype diversity, and recognition of a common mutation, P1173L. Am J Hum Genet 71: 143–153

Authors' address: Th. Müller, MD, Department of Neurology, St. Josef Hospital, Ruhr University Bochum, Gudrunstrasse 56, D-44791 Bochum, Germany, e-mail: thomas.mueller@ruhr-uni-bochum.de

J Neural Transm (2004) [Suppl] 68: 111–116

Coenzyme Q_{10} serum levels in Huntington's disease

J. Andrich[1], **C. Saft**[1], **M. Gerlach**[1,2], **B. Schneider**[1], **A. Arz**[1], **W. Kuhn**[1,3], and **Th. Müller**[1]

[1] Department of Neurology, St. Josef Hospital, Ruhr University, Bochum,
[2] Clinical Neurochemistry, Department for Child and Youth Psychiatry, Würzburg, and
[3] Department of Neurology, Leopoldina Krankenhaus, Schweinfurt, Germany

Summary. Mitochondrial dysfunction contributes to the neurodegenerative process in Huntington's disease (HD). Coenzyme Q_{10} (CoQ_{10}) enhances mitochondrial complex I activity and may therefore provide a therapeutic benefit in HD. We compared serum CoQ_{10} levels of previously untreated – and treated HD patients with those of healthy controls. CoQ_{10} did not significantly (ANCOVA $F_{(dF\,2,\,dF\,55)} = 2.57$; $p = 0.086$) differ between all three groups. However, the post hoc analysis showed no significant ($p = 0.4$) difference between treated HD patients ([CoQ_{10}]: 88.12 [mean] ± 24.44 [SD], [range] 48.75–146.32 [pg/million platelets]) and controls (93.71 ± 20.72, 65.31–157.94), however previously untreated HD patients (70.10 ± 21.12, 38.67–106.14) had marked ($p = 0.051$) lower CoQ_{10} results than treated HD patients and controls ($p = 0.017$). Our results support that CoQ_{10} supplementation in HD patients may reduce impaired mitochondrial function in HD.

Introduction

Huntington's disease (HD) is a neurodegenerative, autosomal dominantly inherited disorder, which manifests in early to middle adulthood (Penney et al., 1997). Characteristic clinical HD symptoms include a wide variety of psychiatric features, cognitive impairment and motor disturbances, which include hyperkinetic syndromes, i.e. chorea or dystonia, which impair initiation, execution and reduction of movements (Furtado et al., 1996; Penney et al., 1997; Paulsen et al., 2001). HD is associated with an expansion of a CAG repeat in the first exon of a gene on chromosome 4. The mutant HD gene encodes an expanded polyglutamine stretch in the N-terminal domain of a protein named huntingtin, which accumulates and causes apoptotic striatal neuronal death (Furtado et al., 1996; Penney et al., 1997; Schilling et al., 1999; Paulsen et al., 2001). The mutant huntingtin increases binding of various proteins, i.e. huntingtin-associated protein (HAP) or GAPDH, all of which are important in cellular energy metabolism. Moreover there is a decreased binding of other proteins, i.e. like huntingtin-interacting protein (HIP) and huntingtin interacting protein–protein interactor (HIPPI), all of which might interact with other important proteins like clathrin and AP2 (Tabrizi et al., 1997; Tabrizi and Schapira, 1999b). Finally, these changes contribute to disturbances of cellular endocytosis and neurotransmitter release in HD (Beal, 1994, 1999). These mechanisms support the neuronal degeneration in specific brain regions, i.e. striatum, frontal cortex, in HD patients (Tabrizi et al., 1997, 1999a, 2000). The slowly evolving neuronal death is accompanied by an impairment of cellular energy metabolism. Accordingly, there were deficiencies of complex II/III and complex IV in post-mortem brain tissue (Tabrizi et al., 1997, 1999b, 2000) and complex I of muscle of HD-patients (Arenas et al., 1998). In Huntington R6/2 transgenic mice, a reduction of aconitase – and striatal complex IV activity appeared (Tabrizi et al., 1999b, 2000). Thus one may assume a mitochondrial dysfunction as contributing pathophysiologic mechanism in HD (Beal, 1999). Administration of Coenzyme Q_{10} (CoQ_{10}, ubiquinone), which is the electron acceptor for mitochondrial complexes I and II and a powerful antioxidant, enhances complex I activity and may therefore provide a certain benefit in neurodegeneration (Cooper and Schapira, 1997; Shults et al., 1997, 2002; Matthews et al., 1998; Müller et al., 2003). Several lines of evidence suggest, that CoQ substitution might slow HD progression according to i.e. animal studies in the transgenic R6/2 mouse model (Keene et al., 2001; Kasparova et al., 2002; Ferrante et al., 2002). However, the corresponding clinical trial in HD patients showed no progression delay in HD under long-term CoQ_{10} supplementation (Kieburtz et al., 2001). Since clinical data on CoQ_{10} metabolism in HD are not available, we compared serum CoQ_{10} levels in previously untreated-, treated HD patients and healthy controls.

Patients and methods

We obtained blood samples drawn from 33 treated (defined as: long-term intake of centrally acting compounds, i.e. antidepressants, neuroleptics, benzodiazepines, tetrabenazine, anticholinergics) HD patients, 8 previously untreated HD subjects and 19 healthy controls (Table 1). Molecular analysis of the CAG expansion was performed, as described (Penney et al., 1997). We scored all HD subjects with the Unified Huntington's Disease Rating Scale (UHDRS) (Huntington Study

Table 1. Clinical characteristics of study participants

Subjects	Gender	Age	CAG	Age of HD onset	UHDRS
Treated HD	17 men, 16 women	54.55 ± 12.26, 35–82	46.3 ± 4.36, 40–55	41.24 ± 2.03, 23–65	45.56 ± 21.14, 12–96
Untreated HD	5 men, 3 women	51.25 ± 11.36, 36–72 s	45.29 ± 4.23, 40–53	44.88 ± 11.96, 30–71	43.38 ± 14.74, 23–73
Controls	10 men, 9 women	56.05 ± 13.85, 31–76			

Data are given as mean $\pm$ standard deviation, range, age (of HD onset) is given in years, *CAG* CAG repeat length, *UHDRS* Unified Huntington's Disease Rating Scale

Group, 1996). Exclusion criteria were supplementation with CoQ_{10}, vitamins, creatine, cholesterol increase or intake of cholesterol modifying compounds (Overvad et al., 1999). We took blood specimen in the morning between 8 a.m. and 9 a.m. after 10 hours fasting from a peripheral vein, drop-wise in plastic vacuum tube. Samples were centrifuged 15 min, 300 × g at 10°C without brake within 10 minutes. We decanted the resulting supernatant and stored it at −80°C. The time period between freezing and analysis of the samples was no longer than three months. There was no light exposure during the whole process. Serum levels of total CoQ_{10} (oxidized CoQ_{10} and reduced $CoQH_2$) were determined by high performance liquid chromatography (HPLC) with electrochemical detection (Götz et al., 2000).

Ethics

All participants gave written informed consent. The study was approved by the ethics committee of the university.

Statistics

Data showed a normal distribution according to the Kolmogorow-Smirnow test. As a result, we only performed parametric tests. We used ANCOVA and set age, sex, number of platelets as covariates. We employed the LSD test for the post hoc analysis and used linear regression for correlation analysis.

Results

CoQ_{10} did not significantly (ANCOVA $F_{(dF\,2,\,dF\,55)} = 2.57$; $p = 0.086$) differ between all three groups. The post hoc analysis of the CoQ_{10} values showed no significant ($p = 0.4$) difference between HD patients ([CoQ_{10}] 88.12 [mean] ± 24.44 [SD], range 48.75–146.32 pg/million platelets) and controls (93.71 ± 20.72, range 65.31–157.94), however previously untreated HD patients (70.10 ± 21.12, 38.67–106.14) had significant ($p = 0.017$) lower CoQ_{10} concentrations than treated ones. The comparison between the controls and the previously treated HD individuals was not significant ($p = 0.051$). There was no significant impact of all set covariates. There were no significant outcomes of the correlation analysis (results not shown).

Discussion

We found no distinct differences of CoQ_{10} levels between HD patients and controls. The found decrease of CoQ_{10} in the previously untreated HD subjects is preliminary result and suffers from the small number of participants. Impact of the chronic drug regime may hypothetically contribute to this outcome. Experimental evidence suggests that neuroleptics or levodopa inhibit complex I in rat brain (Przedborski et al., 1993; Maurer and Moller, 1997). As a result, a shift to an endogenous upregulation of CoQ_{10} synthesis may take place in the more advanced treated HD patients, whereas a putative increased demand for CoQ_{10} hypothetically contributes to the low CoQ_{10} levels in the previously untreated HD patients, who represent the natural course of HD. From this point of view, our study confirms the assumed impaired mitochondrial function as one of the pathophysiologic features of HD (Tabrizi et al., 1997, 1999a, b; Schilling et al., 1999). Moreover we provide circumstantial support for therapeutic approaches with compounds, i.e. creatine, which reduce the mitochondrial energy deficit and the toxicity of free radicals on the mitochondrial complex activity (Schilling et al., 1999; Tabrizi et al., 2002, 2003). However further future more detailed studies on CoQ_{10} levels are necessary for further evaluation of the impact of long-term drug intake under clinical conditions in HD patients, since chronic exposure to environmental toxins, i.e. smoking, also influence the CoQ appearance in peripheral tissue (Smith et al., 1993; Maurer et al., 1997). The results may hopefully contribute to a better understanding of the disappointing outcome of the CoQ_{10} trial in HD (Kieburtz et al., 2001), since studies in patients with Parkinson's disease provided more promising results in addition to the found benefit of CoQ_{10} supplementation in preclinical research in HD models or clinical findings, who indicate a disturbed energy metabolism in HD (Gu et al., 1996; Koroshetz et al., 1997; Kasparova et al., 2002; Shults et al., 2002; Müller et al., 2003).

In conclusion, our study results indicate to a certain extent, that a decrease of CoQ_{10} levels may exist in particular in previously untreated HD patients, which supports the role of mitochondrial dysfunction in the pathophysiology of HD.

Acknowledgement

The authors thank Mrs. S. Poi for her excellent technical assistance.

References

Arenas J, Campos Y, Ribacoba R, Martin MA, Rubio JC, Ablanedo P, Cabello A (1998) Complex I defect in muscle from patients with Huntington's disease. Ann Neurol 43: 397–400

Beal MF (1994) Neurochemistry and toxin models in Huntingtons disease. Curr Opin Neurol 7: 542–547

Beal MF (1999) Mitochondria, NO and neurodegeneration. Biochem Soc Symp 66: 43–54

Cooper JM, Schapira AH (1997) Mitochondrial dysfunction in neurodegeneration. J Bioenerg Biomembr 29: 175–183

Ferrante RJ, Andreassen OA, Dedeoglu A, Ferrante KL, Jenkins BG, Hersch SM, Beal MF (2002) Therapeutic effects of coenzyme Q(10) and remacemide in transgenic mouse models of Huntington's disease. J Neurosci 22: 1592–1599

Furtado S, Suchowersky O, Rewcastle B, Graham L, Klimek ML, Garber A (1996) Relationship between trinucleotide repeats and neuropathological changes in Huntington's disease. Ann Neurol 39: 132–136

Götz ME, Gerstner A, Harth R, Dirr A, Janetzky B, Kuhn W, Riederer P, Gerlach M (2000) Altered redox state of platelet coenzyme Q(10) in Parkinson's disease. J Neural Transm 107: 41–48

Gu M, Gash MT, Mann VM, JavoyAgid F, Cooper JM, Schapira AHV (1996) Mitochondrial defect in Huntington's disease on caudate nucleus. Ann Neurol 39: 385–389

Huntington Study Group (1996) Unified Huntington's Disease Rating Scale: reliability and consistency. Mov Disord 11: 136–142

Kasparova S, Sumbalova Z, Kucharska J, Bystricky P, Mlynarik V, Gvozdjakova A (2002) Effect of coenzyme Q(10) on energy metabolism in aged rats with Huntington's disease. J Mol Cell Cardiol 34: A84

Keene CD, Rodrigues CMP, Eich T, Linehan-Stieers C, Abt A, Kren BT, Steer CJ, Low WC (2001) A bile acid protects against motor and cognitive deficits and reduces striatal degeneration in the 3-nitropropionic acid model of Huntington's disease. Exp Neurol 171: 351–360

Kieburtz K, Koroshetz W, McDermott M, Beal MF, Greenamyre JT, Ross CA, Shoulson I, Cudkowicz ME, Sexton P, Skeuse C, Rosas D, Jenkins B, Rosenblatt A, Sherr M, Hersch S, Cellar J, Guttman M, Brown C, Como PG, Marshall F, DeMarcaida JA, Zimmerman C, Marder K, Moskowitz C, Polanco C, Beltre J, Shannon K, Jaglin JA, Niederman F, Ashizawa T, Hunter C, Hunter C, Hauser R, Walker A, Gauger L, Hahn MA, Delgado H, Dawson T, Siemers E, Seeberger L, Dingmann C, Dubinsky R, Gray C, Weiner WJ, Rodriguez-Batemen D, Albin RL, Wernette K, Harrison M, Rost-Ruffner E, Paulsen J, Rodnitzky RL, Dobson J, Vining L, Suchowersky O, Pantella C, Saint-Hilaire MH, Furtado J, Martin W, Wieler M, Walker F, Hunt V, Raymond L, Almqvist E, Rubin A, Blair B, Marek K, Caplan K, Baker D, Brocht A, Casaceli C, Orme C, Rudolph A, Rumfola L, Schifitto G, Shinaman A, Sulimowicz K, Watts A, Eapen S, Zhang L (2001) A randomized, placebo-controlled trial of coenzyme Q(10) and remacemide in Huntington's disease. Neurology 57: 397–404

Koroshetz WJ, Jenkins BG, Rosen BR, Beal MF (1997) Energy metabolism defects in Huntington's disease and effects of coenzyme Q(10). Ann Neurol 41: 160–165

Matthews RT, Yang L, Browne S, Baik M, Beal MF (1998) Coenzyme Q10 administration increases brain mitochondrial concentrations and exerts neuroprotective effects. Proc Natl Acad Sci USA 95: 8892–8897

Maurer I, Moller HJ (1997) Inhibition of complex I by neuroleptics in normal human brain cortex parallels the extrapyramidal toxicity of neuroleptics. Mol Cell Biochem 174: 255–259

Müller T, Büttner T, Gholipour AF, Kuhn W (2003) Coenzyme Q(10) supplementation provides mild symptomatic benefit in patients with Parkinson's disease. Neurosci Lett 341: 201–204

Overvad K, Diamant B, Holm L, Holmer G, Mortensen SA, Stender S (1999) Coenzyme Q10 in health and disease. Eur J Clin Nutr 53: 764–770

Paulsen JS, Zhao H, Stout JC, Brinkman RR, Guttman M, Ross CA, Como P, Manning C, Hayden MR, Shoulson I (2001) Clinical markers of early disease in persons near onset of Huntington's disease. Neurology 57: 658–662

Penney Jr JB, Vonsattel JP, MacDonald ME, Gusella JF, Myers RH (1997) CAG repeat number governs the development rate of pathology in Huntington's disease. Ann Neurol 41: 689–692

Przedborski S, Jackson-Lewis V, Muthane U, Jiang H, Ferreira M, Naini AB, Fahn S (1993) Chronic levodopa administration alters cerebral mitochondrial respiratory chain activity. Ann Neurol 34: 715–723

Schilling G, Becher MW, Sharp AH, Jinnah HA, Duan K, Kotzuk JA, Slunt HH, Ratovitski T, Cooper JK, Jenkins NA, Copeland NG, Price DL, Ross CA, Borchelt DR (1999) Intranuclear inclusions and neuritic aggregates in transgenic mice expressing a mutant N-terminal fragment of huntingtin. Hum Mol Genet 8: 397–407

Shults CW, Haas RH, Passov D, Beal MF (1997) Coenzyme Q10 levels correlate with the activities of complexes I and II/III in mitochondria from parkinsonian and nonparkinsonian subjects. Ann Neurol 42: 261–264

Shults CW, Oakes D, Kieburtz K, Beal MF, Haas R, Plumb S, Juncos BL, Nutt J, Shoulson I, Carter J, Kompoliti K, Perlmutter JS, Reich S, Stern M, Watts RL, Kurlan R, Molho E, Harrison M, Lew M (2002) Effects of coenzyme Q(10) in early Parkinson disease – Evidence of slowing of the functional decline. Arch Neurol 59: 1541–1550

Smith PR, Cooper JM, Govan GG, Harding AE, Schapira AH (1993) Smoking and mitochondrial function: a model for environmental toxins. Q J Med 86: 657–660

Tabrizi SJ, Schapira AHV (1999b) Secondary abnormalities of mitochondrial DNA associated with neurodegeneration. Biochem Soc Symp 66: 99–110

Tabrizi SJ, Cooper JM, Xuereb JH, Schapira AHV (1997) Mechanisms involved in the pathogenesis of Huntington's disease. Ann Neurol 42: T202

Tabrizi SJ, Cleeter MWJ, Xuereb J, Taanman JW, Cooper JM, Schapira AHV (1999a) Biochemical abnormalities and excitotoxicity in Huntington's disease brain. Ann Neurol 45: 25–32

Tabrizi SJ, Workman J, Hart PE, Mangiarini L, Mahal A, Bates G, Cooper JM, Schapira AHV (2000) Mitochondrial dysfunction and free radical damage in the Huntington R6/2 transgenic mouse. Ann Neurol 47: 80–86

Tabrizi SJ, Rajagopalan B, Styles P, Manners D, Schapira AH, Warner TT (2002) Creatine therapy for Huntington's disease (HD): clinical and-(31)phosphorous magnetic resonance spectroscopy (P-31 MRS) findings in a one year pilot. Mov Disord 17: S319

Tabrizi SJ, Blamire AM, Manners DN, Rajagopalan B, Styles P, Schapira AHV, Warner TT (2003) Creatine therapy for Huntington's disease: clinical and MRS findings in a 1-year pilot study. Neurology 61: 141–142

Authors' address: Prof. Dr. Th. Müller, Department of Neurology, St. Josef Hospital, Ruhr University Bochum, Gudrunstrasse 56, D-44791 Bochum, Germany, e-mail: thomas.mueller@ruhr-uni-bochum.de

J Neural Transm (2004) [Suppl] 68: 117–122

The N-methyl-D-aspartate antagonist memantine retards progression of Huntington's disease

A. Beister[1], **P. Kraus**[2], **W. Kuhn**[2], **M. Dose**[1], **A. Weindl**[3], and **M. Gerlach**[4]

[1] District Hospital, Taufkirchen/Vils,
[2] University Neurological Clinic, St. Josef Hospital, Ruhr University, Bochum,
[3] Neurological Clinic, Technical University, Munich, and
[4] Clinical Neurochemistry, Clinic for Child and Youth Psychiatry, Julius-Maximilians-University, Würzburg, Germany

Summary. According to the excitotoxicity hypothesis, neurotoxicity due to glutamate is regarded as potential factor in the progredient neurodegeneration of Huntington's disease (HD). Memantine, as a glutamate receptor antagonist, should counteract this mechanism. Its effectiveness (up to 30 mg/day) with regard to retardation of progression was thus examined in 27 HD patients in a two year, open and multicentre trial. The results suggest that memantine treatment of HD may be useful in terms of retardation of the progression of the disorder.

Introduction

Huntington's disease (HD) is an inherited (autosomal dominant), progressive neurodegenerative disease characterized by involuntary motor abnormalities, decline in cognitive functions, behavioural abnormalities and loss of capacity for normal tasks of daily living. Nerve cell degeneration is most prominent in the neostriatum (nucleus caudatus, putamen), with lesser losses in the cerebral cortex, hippocampus, thalamus und cerebellum (Lange, 1983).

An unstable CAG-trinucleotide repeat in the huntingtin gene (IT15) on the short arm of chromosome 4 is responsible for the disorder. Despite knowledge regarding the gene location and the polyglutamine (huntingtin) for which it codes (MacDonald and Gusella, 1996), the normal function of this protein remains unclear nor has the role of the mutant huntingtin protein in the pathomechanism of progressive neurodegeneration been elucidated; the abnormal protein may exert an aberrant, toxic function ("toxic gain-of-function"). Altered folding and solubility of the polyglutamine and abnormal interactions with other proteins may also be involved (Jellinger, 2003). Hypotheses regarding excitotoxicity, oxidative stress, compromised energy economy and apoptosis have also been regard as potential paths to neurodegeneration (Beal, 1995;

Gerlach et al., 1996) leading to clinical studies with ketamine (Murman et al., 1997), lamotrigine (Kremer et al., 1999), remacemide and coenzyme Q(10) (Kieburtz et al., 2001).

The excitotoxicity hypothesis is based on observations in animal experiments that intrastriatal injection of quinolinic acid, an endogenous glutamate analogue, elicits neuropathologic changes similar to those in HD (Beal et al., 1986). Increased rate of glutamate synthesis and/or increased sensitivity of cortical glutamatergic projections neurons have thus been proposed to underlie the symptoms of the disorder. Excessive stimulation of N-methyl-D-aspartate (NMDA) receptors by glutamate or by NMDA receptor agonists causes a massive influx of calcium into the cell, and the subsequent activation of calcium-dependent enzymes induces mechanisms which lead to cell death (Gerlach et al., 1996). The toxic effects of excitatory amino acids can be inhibited experimentally by the administration of non-competitive NMDA receptor antagonists such as memantine (Pellegrini and Lipton, 1993).

Memantine, a methyl-substituted adamantane derivate, is a clinically well tolerated and, at therapeutic doses, effective non-competitive NMDA receptor antagonist (Parsons et al., 1999). The potential efficacy of memantine in the treatment of patients with clinically diagnosed or genetically ascertained HD was examined in a multicentre, open and prospective trial lasting 24 months. Progression of the disorder was evaluated by means of rating scales for the assessment of hyperkinesia and competence in tasks of daily living, psychometric tests, electrophysiological investigations and videotape recordings. Efficacy of memantine therapy was judged on the basis of alterations in these parameters during the course of the trial and the results compared with literature on the course of HD.

Materials and methods

Recruitment and enrollment

From July 1994 to October 1997, twenty patients at the Neurological Clinic of the Technical University of Munich and seven at the HD Centre (South) of the Taufkirchen/Vils District Hospital were recruited for the study. All subjects satisfied the a priori inclusion criteria (written informed consent; neurological symptoms of HD; positive family history for the disorder or identification of the mentioned CAG repeat; aged 20 to 60 years), and underwent a general physical as well as neurological examinations prior to commencement of the study. Exclusion criteria were the diagnosis of a second neurological disorder, advanced dementia, schizophrenia or affective psychosis (acute or in patient history); indications of increased suicide risk; serious or decompensated cardiovascular, endocrine, renal or hepatic disease; pregnancy, lactation or dubious contraception; treatment with isoniazid, pyridoxine or allopurinol. Further details concerning the subjects are given in Table 1. A control group of appropriate size could not be recruited.

Clinical assessment

Memantine (Akatinol Memantine®, Merz + Co. GmbH&Co., Frankfurt, Germany) was administered in increasing doses, according to the tolerance of the individual patient, to a maximum of 30 mg/day. For assessment of the clinical effectiveness of the therapy, the parameters listed in Table 2 were assessed at 0, 12 and 24 months by the treating physician. Video recordings of 30 s duration were prepared for each patient in various positions (sitting, calculating while sitting,

Table 1. Patient group data (n = 27)

Gender: 15 male, 12 female
Mean age: 39.7 years (range 18–59)
Mean duration of illness: 30.3 months (range 0–80)
Mean CAG repeats: 46.7 (range 39–61)
Mean IQ (multiple choice vocabulary test): 105.2 (range 83–136)

standing), and evaluated independently and without knowledge of when the recordings were made by two neurologists with experience of HD patients (P.K., W.K.); these raters were not involved in other assessments of patient performance. Global chorea was quantified according to a three point scale (1 = slight, 2 = moderate, 3 = severe; possibility of half-points) for each of head, trunk, arms and legs, and mean values calculated.

Stastical analysis

Descriptive analysis of the data (mean, standard deviation) employed the program SPSS (Version 10.0), while temporal changes in individual parameters were assessed for statistical significance by t-tests for dependent samples, using correction according to Bonferroni.

Results

During the 24 month period of memantine treatment, the primary indices did not show any statistical significant variation. No deterioration was detected by psychometric testing. Also somatic evoked potentials did not exhibit any significant change. Detailed results are listed in Table 3.

Assessment of the video recordings revealed no statistically significant increases in overall mean choreiform unrest. Concordance of the assessments of the two raters was high at the commencement and end of the study period

Table 2. Parameters for assessment of clinical effectiveness

Parameters	References
Skala abnormer unwillkürlicher Bewegungen (Abnormal Involuntary Movements Scale, SKAUB)	Guy (1976)
Modified form of the Unified Huntington's Disease Rating Scale (UHDRS)	Huntington Study Group (1996)
Huntington's Disease Activities of Daily Living (HD-ADL)	Bylsma et al. (1993)
Clinical Global Impression (CGI)	Guy and Bonato (1970)
Syndrom-Kurztest (Short Syndrome Test) according to Erzigkeit (SKT)	Erzigkeit (1992)
Kurztest zur allgemeinen Intelligenz (Brief Test of General Intelligence) (KAI)	Lehrl et al. (1980)
Trail-making test (versions A and B)	Raitan (1956)
Somatic evoked potentials of the Nervus medianus (SEP)	
Standardized video recordings	

Table 3. Changes in experimental parameters during 24 months observation period

	Mean ± SD			p	
	0 months	12 months	24 months	0 v. 24 months	12 v. 24 months
Primary indices					
Video rating	0.67 ± 0.65	0.62 ± 0.49	0.59 ± 0.58	0.78	0.66
SKT	11.5 ± 5.7	10.8 ± 5.8	10.1 ± 6.0	0.74	0.73
KAI	92.4 ± 11.1	90.1 ± 11.8	92.7 ± 16.2	0.47	0.46
SEP: Amplitude (μV)	3.0 ± 1.9	2.0 ± 1.1	1.9 ± 1.0	0.08	0.12
SEP: Latency (ms)	21.4 ± 2.6	19.7 ± 1.7	19.7 ± 1.2	0.30	0.59
Secondary indices					
SKAUB	5.7 ± 5.3	5.3 ± 4.7	6.0 ± 5.3	0.33	0.13
HD-ADL	0.58 ± 0.45	0.81 ± 0.62	0.86 ± 0.59	0.04	0.24
UHDRS	4.4 ± 5.2	5.7 ± 5.9	5.8 ± 6.9	0.32	0.32
CGI	3.4 ± 1.3	4.4 ± 0.6	4.4 ± 0.7	0.01	1.0
Trailmaking A	37.6 ± 22.0	34.0 ± 14.0	34.7 ± 19.0	0.23	0.31
Trailmaking B	87.8 ± 67.9	96.3 ± 37.8	72.3 ± 40.6	0.11	0.85

CGI clinical global impression; *HD-ADL* Huntington's disease activities of daily living; *KAI* brief test of general intelligence; *SEP* somatic evoked potentials of the Nervus medianus; *SKAUB* abnormal involuntary movements scale; *SKT* short syndrome test according to Erzigkeit; *UHDRS* modified form of the unified Huntington's disease rating scale

($p = 0.9$), but at 12 months the first rater observed more severe hyperkinesias ($p = 0.08$). Inspection of individual recordings of particular body regions in temporal succession revealed no significant increase in hyperkinesias.

Deteriorations were measured in both the "Clinical Global Impression" (CGI) and the "HD Activities of Daily Living" (HD-ADL) scales. The mean in both scales increased during the first 12 month rather strong, while there was only little increment in the second year. There were no significant increases in choreiform movements according to either the "Skala abnormer unwillkürlicher Bewegungen" (Scale of Abnormal Involuntary Movements) or the "HD Rating Scale".

Discussion

The current results suggest a neuroprotective effect of memantine in HD. Particularly striking was the reduction in progression of various parameters (CGI, HD-ADL, somatic evoked potential (SEP) amplitudes) during the second year of the study, that is, between 12 and 24 months, during which time no significant changes in these parameters occurred. These results may indicate that the neuroprotective effect of memantine treatment is manifested only after extended administration of the agent.

Reduced progression of disorder-specific symptoms during memantine treatment was also noted in comparison with literature reports of untreated HD patients. An increase in "Total Motor Score" of the Unified HD Rating

Scale (UHDRS) of 3.2 points over six months has been reported in untreated HD patients (Huntington Study Group, 1996) corresponding to a 5.3% decline in motor function during this period, or 21.2% over 24 months. In the current study, a decline of only 4.3% over 24 months was measured with a modified form of the UHDRS. The "Total Functional Capacity" (TFC) arm of the "HD Functional Capacity Scale" concurs in general with the HD-ADL employed in the current study. Untreated HD patients exhibited a mean decline in TFC of 0.5 points over six months (Shoulson and Fahn, 1979); during memantine treatment, an increase of 0.28 points on the HD-ADL scale was measured. These results correspond with a 15.4% decline in competence in the tasks of daily living in untreated patients over 24 months, but a 9.3% reduction in progression of incapacity in the memantine-treated patients. Further, amplitude of somatic evoked potentials were reduced by 1.7 μV in untreated HD patients (Ehle et al., 1984) whereas the reduction in memantine-treated patients was 1.1 μV.

In summary, memantine treatment of HD appears to be very promising in terms of retardation of the progression of the disorder. Further, placebo-controlled studies are required to confirm these findings.

Acknowledgements

This study was partially supported by a grant from Merz + Co. GmbH&Co. (Frankfurt, Germany).

References

Beal MF (1995) Ageing, energy and oxidative stress in neurodegenerative diseases. Ann Neurol 38: 357–366

Beal MF, Kowall NW, Ellison DW, Mazurek MF, Swartz KJ, Martin JB (1986) Replication of the neurochemical characteristics of Huntington's disease by quinolinic acid. Nature 321: 29–38

Bylsma FW, Rothlind J, Hall MR, Folstein SE, Brandt J (1993) Assessment of adaptive functioning in Huntington's disease. Mov Disord 8: 183–190

Ehle AL, Stewart RM, Lellied NA, Leventhal NA (1984) Evoked potentials in Huntington's disease. Arch Neurol 41: 379–382

Erzigkeit H (1992) SKT: Ein Kurztest zur Erfassung von Gedächtnis- und Aufmerksamkeitsstörungen. Beltz Test, Weinheim

Gerlach M, Riederer P, Youdim MBH (1996) Molecular mechanisms for neurodegeneration: synergism between reactive oxygen species, calcium and excitotoxic amino acids. Adv Neurol 169: 177–194

Guy W (1976) ECDEU. Assessment manual for psychopharmacology, revised edn. US Department of Health, Education and Welfare, Washington DC

Guy W, Bonato RR (1970) Manual for the ECDEU assessment battery, 2nd revised edn. National Institute of Mental Health, Chevy Chase

Huntington Study Group (1996) Unified Huntington's disease rating scale. Reliability and consistency. Mov Disord 11: 136–142

Jellinger KA (2003) General aspects of neurodegeneration. J Neural Transm [Suppl] 65: 101–144

Kieburtz K, Koroshetz W, McDermott M, Beal MF, Greenamyre JT, Ross CA, Shoulson I, Cudkowicz ME, Sexton P, Skeuse C, Rosas D, Jenkins B, Rosenblatt A, Sherr M, Hersch S, Cellar J, Guttman M, Brown C, Como PG, Marshall F, DeMarcaida JA, Zimmerman C, Marder K, Moskowitz C, Polanco C, Beltre J, Shannon K, Jaglin JA, Niederman F, Ashizawa T, Hunter C, Hauser R, Walker A, Gauger L, Hahn MA, Delgado H, Dawson T, Siemers E, Seeberger L, Dingmann C, Dubinsky R, Gray C, Weiner WJ, Rodriguez-Batemen D,

Albin RL, Wernette K, Harrison M, Rost-Ruffner E, Paulsen J, Rodnitzky RL, Dobson J, Vining L, Suchowersky O, Pantella C, Saint-Hilaire MH, Furtado J, Martin W, Wieler M, Walker F, Hunt V, Raymond L, Almqvist E, Rubin A, Blair B, Marek K, Caplan K, Baker D, Brocht A, Casaceli C, Orme C, Rudolph A, Rumfola L, Schifitto G, Shinaman A, Sulimowicz K, Watts A, Eapen S, Zhang L, The Huntington Study Group (2001) A randomized, placebo-controlled trial of coenzyme Q(10) and remacemide in Huntington's disease. Neurology 57: 397–404

Kremer B, Clark CM, Almqvist EW, Raymond LA, Graf P, Jacova C, Mezei M, Hardy MA, Snow B, Martin W, Hayden MR (1999) Influence of lamotrigine on progression of early Huntington disease – a randomized clinical trial. Neurology 53: 1000–1011

Lange HW (1983) Quantitative changes of telencephalon, diencephalon and mesencephalon in Huntington's chorea, postencephalitic and idiopathic parkinsonism. Verh Anat Ges 75: 923–925

Lehrl S, Gallwitz A, Blaha L, Fischer B (1980) Kurztest für allgemeine Intelligenz KAI. VLESS, Vaterstetten

MacDonald ME, Gusella JF (1996) Huntington's disease: translating a CAG repeat into a pathogenic mechanism. Curr Opin Neurobiol 6: 638–643

Murman DL, Giordani B, Mellow AM, Johanns JR, Little RJA, Hariharan M, Foster NL (1997) Cognitive, behavioral, and motor effects of the NMDA antagonist ketamine in Huntington's disease. Neurology 49: 153–161

Parsons CG, Danysz W, Quack G (1999) Memantine is a clinically well tolerated NMDA receptor antagonist – a review of preclinical data. Neuropharmacology 138: 735–767

Pellegrini JW, Lipton SA (1993) Delayed administration of memantine prevents N-methyl-D-aspartate receptor mediated neurotoxicity. Ann Neurol 33: 403–407

Raitan RM (1956) Trail Making Test. Manual for administration, scoring and interpolation. Indiana University Press, Indianapolis

Shoulson I, Fahn S (1979) Huntington disease; clinical care and evaluation. Neurology 29: 1–3

Authors' address: Prof. Dr. M. Gerlach, Clinical Neurochemistry, Child and Youth Psychiatry and Psychotherapy, Füchsleinstrasse 15, D-97080 Würzburg, Germany, e-mail: manfred.gerlach@mail.uni-wuerzburg.de

J Neural Transm (2004) [Suppl] 68: 123–133

Transcranial sonography in spinocerebellar ataxia type 3

Th. Postert[1,2], J. Eyding[2], D. Berg[3], H. Przuntek[2], G. Becker[4], M. Finger[2], and L. Schöls[2]

[1] Department of Neurology, St. Vincenz Hospital, Paderborn,
[2] Department of Neurology, St. Josef Hospital, Ruhr-University, Bochum,
[3] Department of Human Genetics, University of Tübingen, and
[4] Department of Neurology, University of Homburg, Germany

Summary. *Background*: Transcranial sonography (TCS) identifies hyperechogenic basal ganglia in extrapyramidal disorders such as Parkinson's disease or dystonia and reveals brain atrophy reflecting the stage of degeneration in Huntington's disease. Aim of the present study was to evaluate the diagnostic potential of TCS in spinocerebellar ataxia type 3 (SCA3), a neurodegenerative disease affecting the cerebellum, multiple pontine nuclei, substantia nigra, pallidum, putamen, caudate nucleus and long spinal tracts.

Methods: 15 patients with a molecular diagnosis of SCA3 and 15 age- and sex-matched healthy control individuals were examined with TCS. Echogenicity of the substantia nigra, caudate nucleus, pallidum, putamen, dentate nucleus and cerebellar white matter were determined quantitatively (substantia nigra by measuring the area of incresed echogenicity) or semiquantitatively (0 = none, 1 = mild, 2 = marked) on both sides and compared with the echotexture of defined brain structures. Additionally, the width of the lateral ventricles, the 3rd and 4th ventricle was measured in both groups.

Results: SCA3 patients exhibited hyperechogenicities of the cerebellar white matter (57%), substantia nigra (40%), the dentate nucleus (54%), putamen (40%) and pallidum (40%) significantly more frequently ($p < 0.05$) than controls (20%, 13%, 9%, 0%, and 0% in the corresponding areas). In none of the healthy individuals a marked signal increase could be observed, whereas 53% of SCA3 subjects had at least one region with marked echogenicity. Cerebellar atrophy as demonstrated by an enlarged 4th ventricle was observed in all SCA3 patients whereas this structure could not be differentiated from surrounding parenchyma due to its small size in healthy individuals. 3rd and lateral ventricles were significantly larger in SCA patients as compared to controls ($p < 0.05$).

Conclusions: TCS is a suitable and non-invasive bed-side method to detect basal ganglia hyperechogenic lesions and posterior fossa abnormalities in SCA3 patients. The pattern of echotexture alterations and brain atrophy most likely reflects distribution and morphology of the neurodegenerative process.

Introduction

Transcranial color-coded sonography (TCS) enables not only the two-dimensional visualization of the intracranial arterial vessels but displays brain parenchyma through the intact skull unequivocally as well. Particulary in acute stroke patients this technique allows reliable assessment of intracranial vascular pathology and has replaced CT or MR angiography in many clinics (Gerriets et al., 2000, 2002; Postert et al., 1999a). In addition, it could be shown in numerous studies that transcranial B-mode sonography is useful and sensitive in demonstrating brain atrophy and parenchymal lesions such as brain tumours or intracranial hemorrhages (Becker et al., 1955a; Seidel et al., 1993, 1995). Even more interesting, the finding that TCS displays abnormalities in the echotexture of the basal ganglia in various movement disorders provided new insights in the diagnostic evaluation of these diseases. In Parkinson's disease, for example, an increase of echogenicity of the subtantia nigra has been demonstrated in the majority of patients. It has been speculated that accumulation of heavy metals or changes of the cytoarchitecture and glial-cell composition within this region are possible reasons for this morphological alteration (Becker et al., 1995b; Walter et al., 2002, 2003). Idiopathic dystonia has been shown to be associated with lesions of the lenticular nucleus (Naumann et al., 1996). Patients with Huntington's disease exhibited more often hyperechogenic lesions of the caudate nucleus and to a minor extent of the SN than healthy individuals. Moreover, extent and morphology of brain atrophy could be precisely visualized and closely matched with CT or MRI findings (Postert et al., 1999b). All these studies indicate that TCS may complement CT and MRT investigations in a variety of degenerative CNS disorders. However, patients with spinocerebellar ataxias have not been investigated with TCS hitherto. Aim of the present study was to evaluate basal ganglia morphology as well as cerebral and cerebellar atrophy in patients with spinocerebellar ataxia type 3 (SCA3).

Subjects and methods

Clinical data

The present study included 15 patients with the molecular diagnosis of SCA3 and 15 age- and sex-matched healthy control subjects without a history of neurological diseases. SCA3 patients underwent a standardized clinical examination by one of us (LS). Severity of ataxia was rated using an ataxia score comprising 8 items where 0 points means no ataxia and 40 points represent most severe ataxia (Schulte et al., 2001). Additionally, the presence or absence of extrapyramidal symptoms such as parkinsonism or dystonia and pyramidal signs was documented. Clinical, genetic and demographic data of SCA3 patients are summarized in Table 1. All individuals gave informed consent to participate in the study.

Transcranial sonography

Only patients and volunteers with a sufficient acoustic bone window were included in this study. All subjects were examined using a phased array ultrasound system (Siemens Sonoline Elegra, Siemens Medical Solutions, Inc., Ultrasound Group, Issaquah, WA, USA) equipped with a 2.5 MHz 90° sector scan. All TCS examinations were performed by three experienced ultrasound examiners (TP, JE, DB).

Table 1. Clinical and demographic characteristics in 15 SCA3 patients

	SCA3 patients
Number of individuals	15
Male:female	10:5
Age	48.5 ± 12.5 years
CAG repeats (n = 14)	71.7 ± 3.2
Ataxia score according to (12)	17.3 ± 7.8 points
Duration of the disease	12.7 ± 7.2 years

Using the preauricular acoustic bone window, a B-mode image was created on both sides, depicting brain parenchyma in different shades of grey. The penetration depth was set at 14 to 16 cm, the dynamic range was 45 dB. Firstly, the butterfly shaped brainstem was visualized in the axial untilted sections. The substantia nigra was identified within the structure of the mesencephalic brainstem as clearly as possible and quantified by encircling and measuring the area of hyperechogenic signals as previously described (Berg et al., 2001). Echogenicity of the brainstem raphe was rated semiquantitatively on a three point scale: 0 = raphe not visible/isoechogenic raphe compared with adjacent brainstem parenchyma, 1 = slightly echogenic raphe, 2 = normal ramal raphe echogenicity. Then the scanning plane was shifted cranially to visualise diencephalic structures. Basal ganglia were best illustrated from the contralateral probe position. The echogenicity of the basal ganglia in this scanning plane was assessed semiquantitatively using a three point scale: 0 = normal (isoechogenicity compared with normal white matter), 1 = mild hyperechogenicity, 2 = marked hyperechogenicity. The following brain areas were analysed according to this rating scale: caudate nucleus, pallidum and putamen. Echogenicity and echotexture of each

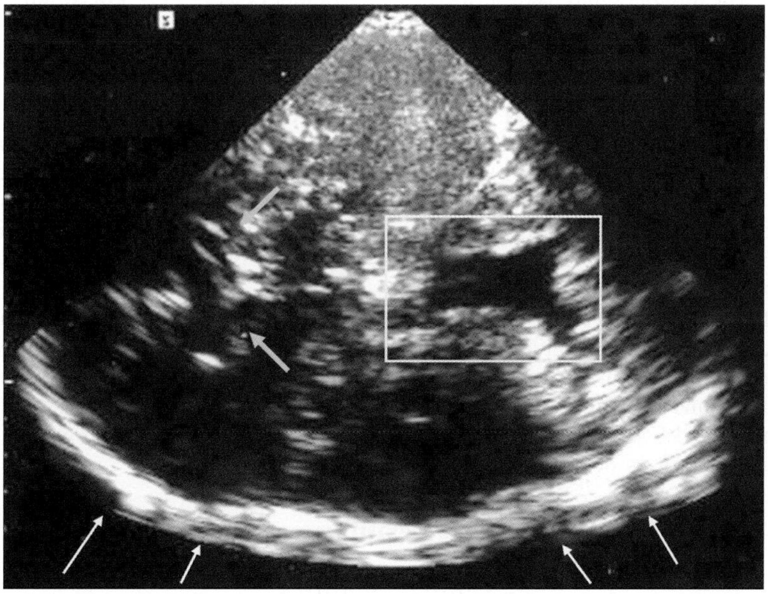

Fig. 1. Axial mesencephalic plane of section in a SCA3 patient: left side = frontal pole, right side = occipital pole; small arrows = contralateral skull (examination depth 15 cm); bold arrows = frontal horns of lateral ventricles; rectangle contains butterfly-shaped mesencephalic brainstem, enlarged 4th ventricle and hyperechogenic dentate nuclei (for enlarged details of the rectangle see Fig. 2)

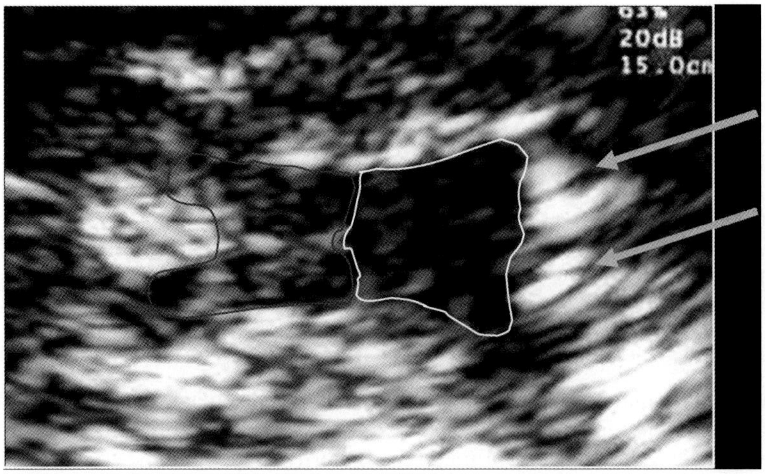

Fig. 2. Enlarged details in the same patient as Fig. 1. The plane of section is approximately corresponding to the rectangle of Fig. 1, the probe has been slightly tilted into the caudal direction in order to improve visualization of the dentate nuclei: butterfly shaped mesencephalic brainstem (grey line), 4th ventricle as a hypoechogenic circular structure in immediate dorsal vicinity of the mesencephalic brainstem (white line), dorsal to the 4th ventricle the dentate nuclei are displayed as hyperechogenic structures (arrows)

basal ganglia region was compared with the echomorphology of the surrounding brain parenchyma and the echogenicity of the corresponding contralateral structures. In addition, the size of the anechogenic ventricular system was evaluated. The maximal size of the frontal horns (reaching from the septum pellucidum to the tip of the frontal horn) and the third ventricle were measured in standardized scanning planes as described earlier (Seidel et al., 1995). For all measurements of ventricular width, hyperechogenic borders of the ependyma were included. Finally, the dorsal parts of the probe were tilted caudally in order to visualize the dentate nucleus, the 4th ventricle and the cerebellar hemispheres. The 4th ventricle can be identified as a hypo- to anechogenic circular structure in immediate dorsal vicinity of the mesencephalic brainstem. In case of missing cerebellar atrophy this structure cannot be displayed accurately. The dentate nucleus can be identified bilaterally adjacent to the hypoechogenic 4th ventricle (see also Figs. 1 and 2). Echogenicity of the dentate nucleus and the cerebellum were determined semiquantitatively as described above. Size of the fourth ventricle was measured by encircling this structure or by measuring the maximal axial diameter depending on which parameter could be visualised with higher accuracy.

Width of the lateral, 3rd and 4th ventricles were compared using Wilcoxon's matched pairs test. The frequency of basal ganglia hyperechogenicities in SCA3 patients and healthy individuals were compared with the chi-square test (asymptotic two-sided significance). Correlation of width of the lateral, 3rd and 4th ventricles as well as hyperechogenic areas of the substantia nigra with age, age at onset and CAG repeat length were performed using Pearson's correlation coefficient. Level of significance was determined as 0.05 (two sided).

Results

Clinical findings

Disease duration in SCA3 patients was between 2 and 23 years. Ataxia scores ranged from 9 to 34 points. Four patients had additional extrapyramidal signs (3 parkinsonian symptoms, 1 dystonia) and 10 patients presented with spasticity in addition to ataxia (Table 1).

Table 2. Parameters of brain atrophy and echogenicity of the substantia nigra in 15 SCA patients and 15 age and sex-matched healthy individuals

	SCA3 patients	Healthy individuals	Level of significance
Area of substantia nigra (both sides)	$0.21 \pm 0.1\,cm^2$	$0.13 \pm 0.05\,cm^2$	$P < 0.0005$
Width of 3rd ventricle	4.8 ± 1.6 mm	3.5 ± 0.6 mm	$P < 0.05$
Width of lateral ventricles (both sides)	18.5 ± 3.2 mm	16.4 ± 1.22 mm	$P < 0.005$
Axial diameter of 4th ventricle ($n = 9$)	16.9 ± 2.7 mm	n.a.	
Circumference of 4th ventricle ($n = 5$)	68.6 ± 13.5 mm	n.a.	

n.a. not applicable, all values in Tables 1 and 2 are mean values ± one standard deviation

Transcranial sonography

General

In two patients the substantia nigra and in one other the dentate nucleus and 4th ventricle could not be identified due to partially insufficient bone windows. In none of the healthy individuals the 4th ventricle could be clearly identified. This was most likely due to the small size of this structure in individuals without cerebellar atrophy.

Brain atrophy

The width of the 3rd ventricle and of the lateral ventricles could be measured in all but one SCA3 patient. Mean values of all ventricles were significantly larger in patients as compared to healthy individuals (see Table 2). In all SCA patients the 4th ventricle could be easily identified as a hypo- to anechogenic structure dorsally to the mesencephalic brainstem (Figs. 1 and 2), whereas this was not possible in healthy individuals. Diameters and circumferences of the 4th ventricle varied between patients (see Table 2) depending on the extent of cerebellar atrophy.

Basal ganglia and other parenchymal alterations

Patterns of hyperechogenic lesions and genetical findings in all 15 SCA3 individuals are summarized in Table 3.

Substantia nigra, raphe, caudate nucleus, pallidum and putamen

Mean echogenic area of the substantia nigra was significantly higher in SCA3 patients as compared to healthy controls ($p < 0.05$; see Table 2). In 11 out of 15 SCA3 patients the area of substantia nigra hyperechogenicity exceeded $0.2\,cm^2$ and in 6 patients it was larger than $0.25\,cm^2$ at least on one side (Fig. 3). According to the literature a hyperechogenic area of more than $0.25\,cm^2$ is above the 90th percentile of the healthy population and for this reason considered to be pathological (Berg et al., 1999a). In only 1 out of 3 patients with additional parkinsonian symptoms the substantia nigra exhibited a substantial

Table 3. Clinical, genetic, and sonographic findings in 15 SCA3 patients

Initials	Age, gender	Disease duration (years)	CAG repeats	Ataxia score	Add. symptoms	SN right (cm^2)	SN left (cm^2)	Put. right	Put. left	Pall. right	Pall. left	Raphe	Dentate nuclei	4th v. diam (mm)	4th v. circ. (mm)	Cereb. wm
BC	35, f	5	75	8	S	0.15	0.22	0	0	0	0	2	2	17		0
GG	59, m	18	70	30	P	0.33	0.28	0	1	1	0	2		16		0
HL	56, m	23	71		P	0.17	0.21	0	0	0	0	2	2	14		0
HS	40, m	13	73	11	S	0.2	0.22	0	0	0	0	2	2	14		1
KT	39, m	9	73	16	S	0.12	0.21	0	0	0	0	2	1	19		1
BT	58, f	22	69	34	S;P	0.12	0.1	0	0	0	0	2		14		0
MR	39, m	10	72	10	S	0.08	0.09	0	0	0	0	2	2	17		1
KR	74, m	19	65	19		0.16	0.13	0	0	0	0	2	2	19		1
IL	63, f	6		14		0.21	0.28	0	0	0	0	2	0	33		1
MS	39, m	17	76	22	S	0.41	0.27	1	1	1	0	2	1		71	1
FS	32, m	2	74	12	S		0.35	1	1	2	1	2				
DK	54, m	15	72	17	S		0.21	0	1	1	2	2	0		46	0
OO	35, m	2	72	9	S	0.3	0.3	2	2	1	2	1	0		76	2
KO	57, f	8	67	16		0.33	0.31	2	2	0	1	2	0		69	1
EB	47, f	21	76	24	D;S	0.07	0.1	0	0	0	0	2	0		81	0

m male, *f* female; *SN* substantia nigra; *4th v.* fourth ventricle, *diam.* axial diameter, *circ.* circumference; *0,1 or 2* no, mild or marked hyperechogenicity; *cereb. wm* cerebellar white matter; *put.* putamen, *pall.* pallidum; *add. symptoms* additional symptoms, *S* spasticity, *D* dystonia, *P* parkinsonism

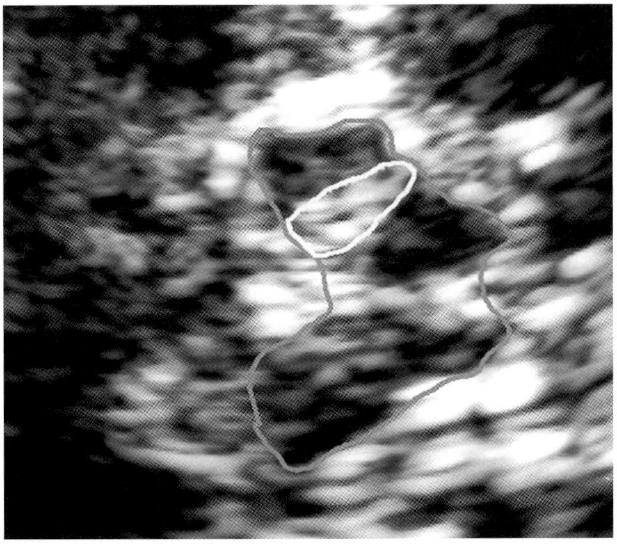

Fig. 3. Mesencephalic brainstem (grey line) with a marked hyperechogenicity of the substantia nigra (white line)

increase in the echotexture, in the other two it was normal. The raphe echogenicity was reduced in one SCA3 patient and in no healthy individual.

In 6 SCA3 patients and 1 healthy individual hyperechogenic areas in the putamen or pallidum were detected ($p<0.05$), whereas the caudate nucleus never exhibited such an alteration of echotexture. 5 out of 6 patients had bilateral hyperechogenic areas of the pallidum and putamen, in one patient there was a mild hyperechogenicity on one side of the putamen and the pallidum was hyperechogenic on both sides. 3 patients had a marked hyperechogenicity at least in one basal ganglia area.

Dentate nucleus and cerebellum

The dentate nucleus exhibited a marked hyperechogenicity in 5 and a mild increase of echotexture in 2 SCA3 patients (Fig. 2). This was not observed in the control subjects ($p<0.05$). In 7 patients and 3 healthy individuals the cerebellar white matter was mildly hyperechogenic, in one patient there was a marked signal increase.

Correlation of TCS findings with clinical and genetic data

In SCA3 patients age and age of onset correlated inversely with CAG repeat length confirming earlier results (correlation $= -0.809$, $p<0.001$, and -0.924, $p<0.01$, respectively) (Schöls et al., 1996). CAG repeat length did not significantly influence the echogenic area of the substantia nigra nor atrophy measured as width of the lateral, 3rd and 4th ventricle. Similarly, echogenic area of the substantia nigra and width of the lateral, 3rd and 4th ventricle did not depend on age or age of onset, apart from correlation of width of 4th ventricle and age of onset (0.706, $p<0.033$). Patients with extrapyramidal signs did not show different echogenicity of the substantia nigra than individuals without extrapyramidal features. Neither substantia nigra nor putamen nor pallidum echogenicity corresponded to parkinsonian symptoms. In three patients Parkinson syndrome of akinetic rigid type was the major manifestation of SCA3. Levodopa response was highly variable. Whereas patient HL (Table 3) had substantial benefit from levodopa treatment for more than 10 years this was limited to 2–3 years in patient BT, and no clear levodopa response was seen in patient GG. Patients BT and GG but not HL experienced psychotic reactions when treated with a daily dose of 500 mg levodopa. Levodopa response did not correlate with TCS findings (Table 3). Ataxia scores in patients with marked hyperechogenicities of the dentate nucleus did not differ from subjects with normal echogenicity.

Discussion

In the present study TCS demonstrated hyperechogenic lesions of the basal ganglia and cerebellum in approximately 50% of patients with SCA3. The dentate nucleus, cerebellar white matter and substantia nigra were the regions most commonly affected. Furthermore, TCS depicted a pattern of generalized brain atrophy with accentuation of posterior fossa structures.

In Parkinson's disease hyperechogenicities of the SN were identified in the vast majority of affected individuals (Walter et al., 2002, 2003; Becker and Berg, 2001). Furthermore, the age of disease onset correlated inversely with the area of echogenicity and the area of increased echotexture was larger contralateral to the side with more severe symptoms (Berg et al., 2001). More recently, the rate of increased echogenicity in a large population of healthy individuals was found to be 8.6%. Since this echopattern was associated with a marked decrease in the accumulation of (18-F)-dopa in the caudate nucleus and putamen, it was assumed that a functional impairment indicating an increased vulnerability of the nigrostriatal system may cause these alterations (Berg et al., 1999a). Explanations of the underlying pathophysiology of hyperechogenic lesions include iron accumulation or deposition of other heavy metals and focal gliosis (Berg et al., 1999).

In comparison to TCS studies in other basal ganglia disorders our study demonstrated a pattern of hyperechogenic lesions and brain atrophy in SCA3 patients that corresponds to the pathological findings of the limited number of necropsy and MRI studies published hitherto. SCA3 is supposed to be pathologically different from other SCA subtypes like SCA1, SCA2 and SCA6 in that Purkinje cells are almost preserved whereas pronounced degeneration is seen in the dentate nucleus (Woods and Schaumburg, 1972; Rosenberg et al., 1976; Dürr et al., 1996; Takiyama et al., 1994; Sequeiros and Coutinho, 1993). In contrast to SCA1 and SCA2, in SCA3 the basal ganglia are often affected by the disease. Neuropathological studies have depicted neuronal loss in the striatum and the substantia nigra (Dürr et al., 1996). Accordingly, extrapyramidal motor signs, although not specific for any SCA subtype, frequently occur in SCA3 patients. Even a levodopa-responsive variant with parkinsonism at the beginning of the disease has been described in SCA3 (Schöls et al., 2000).

These findings were confirmed in recently published MR volumetric studies that demonstrated reduction of the caudate and putamen volumes in SCA3 patients compared to controls (Klockgether et al., 1998). TCS findings in our study underline the variability of pathomorphological findings in SCA3 patients. We observed hyperechogenic lesions not only in the dentate nucleus but additionally in the substantia nigra and the pallidum which were hyperechogenic in approximately 50% of SCA3 patients. Correlation of TCS signal alteration in basal ganglia, parkinsonian symptoms and L-Dopa response needs further evaluation in larger groups of patients. Additionally, functional imaging studies are warranted to determine pre- and postsynaptic dopaminergic abnormalities and their correlation to TCS findings.

MRI studies revealed enlargement of the 4th ventricle to be the most characteristic finding in SCA3 (Schöls et al., 1997). The clearly enlarged 4th ventricles depicted by TCS in all SCA3 patients demonstrated good correspondence in morphometric analyses. We could not establish a correlation of CAG repeat length or age with TCS findings. However, this may be due to small numbers.

TCS findings in SCA3 patients were not specific for the disease in that none of the hyperechogenicities found in different brain areas alone is capable to confirm the diagnosis of SCA3 in a suspective patient. However, the combination of hyperechogenicites in various areas of the basal ganglia and the cerebellum associated with cerebellar atrophy may give a strong indication towards this diagnosis. Up to now, all routinely used structural neuroimaging methods can only give limited information concerning the diagnosis of SCA3 (Schöls et al., 1997). Using MRI it is mainly cerebellar and pontine atrophy, enlargement of the fourth ventricle and a hyperdensity of the dentate nuclei that confirms the diagnosis. Therefore, TCS may be used to add further complementory information to other neuroimaging data. As TCS is broadly available, easy to perform, cheap and without the disadvantage of side effects like radiation it may indeed be regarded as an interesting additional diagnostic tool for the diagnosis of SCA3.

In summary our data indicate that underlying pathological processes in SCA3 are reflected in alterations of the brain echotexture in TCS examinations. Although presumably not specific for a single type of autosomal dominant ataxia, this pattern of alterations has not been observed in other basal ganglia diseases such as Parkinson's disease. The present study firstly describes sonographic findings in genetically determined ataxias. Future studies will have to clarify the sonographic appearance of other inherited cerebellar diseases in comparison to SCA3 and should compare these findings with MRI studies.

References

Becker G, Berg D (2001) Neuroimaging in basal ganglia disorders: perspectives for transcranial ultrasound. Mov Disord 16: 23–32

Becker G, Krone A, Schmitt K, Woydt M, Hofmann E, Lindner A, Bogdahn U, Gahn G, Roosen K (1995a) Preoperative and postoperative follow-up in high-grade gliomas: comparison of transcranial color-coded real-time sonography and computed tomography findings. Ultrasound Med Biol 9: 1123–1135

Becker G, Seufert J, Bogdahn U, Reichmann H, Reiners K (1995b) Degeneration of substantia nigra in chronic Parkinson's disease visualized by transcranial color-coded real-time sonography. Neurology 45: 182–184

Berg D, Becker G, Zeiler B, Tucha O, Hofmann E, Preier M, Benz P, Jost W, Reiners K, Lange KW (1999a) Vulnerability of the nigrostriatal system as detected by transcranial ultrasound. Neurology 53: 1026–1031

Berg D, Grote C, Rausch W-D, Mäurer M, Wesemann W, Riederer P, Becker G (1999) Iron accumulation in the substantia nigra in rats visualized by ultrasound. Ultrasound Med Biol 25: 901–904

Berg D, Siefker C, Ruprecht-Dörfler P, Becker G (2001) Relationship of substantia nigra echogenicity and motor function in elderly subjects. Neurology 56: 13–17

Dürr A, Stevanin G, Cancel G, Duyckaerts C, Abbas N, Didierjean O (1996) Spinocerebellar ataxia 3 and Machado-Joseph disease: clinical, molecular, and neuropathological features. Ann Neurol 39: 490–499

Gerriets T, Postert T, Goertler M, Stolz E, Schlachetzki F, Sliwka U, Seidel G, Weber S, Kaps M (2000) DIAS I: duplex-sonographic assessment of the cerebrovascular status in acute stroke. A useful tool for future stroke trials. Stroke 31: 2342–2345

Gerriets T, Goertler M, Stolz E, Postert T, Sliwka U, Schlachetzki F, Seidel G, Weber S, Kaps M (2002) Feasibility and validity of transcranial duplex sonography in patients with acute stroke. J Neurol Neurosurg Psychiatry 73: 17–20

Klockgether T, Skalej M, Wedekind D, Luft AR, Welte D, Schulz JB, Abele M, Bürk K, Laccone F, Brice A, Dichgans J (1998) Autosomal dominant cerebellar ataxia type 1: MRI- based volumetry of posterior fossa structures and basal ganglia in spinocerebellar ataxia types 1, 2 and 3. Brain 121: 1687–1693

Naumann N, Becker G, Toyka KV, Supprian T, Reiners K (1996) Lenticular nucleus lesion in idiopathic dystonia detected by transcranial sonography. Neurology 47: 1284–1290

Postert T, Braun B, Meves S, Köster O, Przuntek H, Weber S, Büttner T (1999a) Contrast-enhanced transcranial color-coded sonography in acute hemispheric brain infarction. Stroke 30: 1819–1826

Postert T, Lack B, Kuhn W, Jergas M, Andrich J, Braun B, Przuntek H, Sprengelmeyer R, Agelink M, Büttner T (1999b) Basal ganglia alterations and brain atrophy in Huntington's disease depicted by transcranial real time sonography. J Neurol Neurosurg Psychiatry 67: 457–462

Rosenberg RN, Nyhan WL, Bay C, Shore P (1976) Autosomal dominant striatonigral degeneration. A clinical, pathologic, and biochemical study of a new genetic disorder. Neurology 26: 703–714

Schöls L, Amoiridis G, Epplen JT, Langkafel M, Przuntek H, Riess O (1996) Relations between genotype and phenotype in German patients with the Machado-Joseph disease mutation. J Neurol Neurosurg Psychiatry 61: 466–470

Schöls L, Amoiridis G, Buttner T, Przuntek H, Epplen JT, Riess O (1997) Autosomal dominant cerebellar ataxia: phenotypic differences in genetically defined subtypes? Ann Neurol 42: 924–932

Schöls L, Peters S, Szymanski S, Krüger R, Lange S, Hardt C, Riess O, Przuntek H (2000) Extrapyramidal motor signs in degenerative ataxias. Arch Neurol 57: 1495–1500

Schulte T, Mattern R, Berger K, Szymanski S, Klotz P, Kraus PH, Przuntek H, Schöls L (2001) Double-blind crossover trial of trimethoprim-sulfamethoxazole in spinocerebellar ataxia type 3/Machado-Joseph disease. Arch Neurol 58: 1451–1457

Seidel G, Kaps M, Dorndorf W (1993) Transcranial color-coded duplex sonography of intracerebral hematomas in adults. Stroke 24: 1519–1527

Seidel G, Kaps M, Gerriets T, Hutzelmann A (1995) Evaluation of the ventricular system in adults by transcranial duplex sonography. J Neuroimaging 5: 105–108

Sequeiros J, Coutinho (1993) Epidemiology and clinical aspects of Machado-Joseph disease. In: Harding AE, Deufel T (eds) Advances in neurology, vol 61. Raven Press, New York, pp 139–153

Takiyama Y, Oyanagi S, Kawashima S, Sakamoto H, Saito K, Yoshida M (1994) A clinical and pathological study of a large Japanese family with Machado-Joseph disease tightly linked to the DNA markers on chromosome 14q. Neurology 44: 1302–1308

Walter U, Wistock M, Benecke R, Dressler D (2002) Substantia nigra echogenicity is normal in non-extrapyramidal cerebral disorders but increased in Parkinson's disease. J Neural Transm 109: 191–196

Walter U, Niehaus L, Probst T, Benecke R, Meyer BU, Dressler D (2003) Brain parenchyma sonography discriminates Parkinson's disease and atypical parkinsonian syndromes. Neurology 60: 74–77

Woods PT, Schaumburg HH (1972) Nigro-spino-dentatal degeneration with nuclear ophthalmoplegia. A unique and partially treatable clinico-pathological entity. J Neurol Sci 17: 149–166

Authors' address: Dr. Th. Postert, Department of Neurology, St. Vincenz Hospital, Landeshospital, Kisau 14, D-33098 Paderborn, Germany, e-mail: t.postert@vincenz.de

J Neural Transm (2004) [Suppl] 68: 135–145

Therapeutic strategies in Friedreich's ataxia

L. Schöls[1], **Ch. Meyer**[2], **G. Schmid**[3], **I. Wilhelms**[2], and **H. Przuntek**[2]

[1] Center of Neurology and Hertie-Institute for Clinical Brain Research, University of Tübingen,
[2] Department of Neurology, and
[3] Institute of Radiology, St. Josef Hospital, Ruhr-University, Bochum, Germany

Summary. Friedreich's ataxia is caused by a pronounced lack of frataxin, a mitochondrial protein of not fully understood function. Lack of frataxin homologues in yeast and mice leads to increased sensitivity to oxidative stress, depletion of proteins with iron-sulfur clusters like respiratory chain complexes I–III and aconitase, and to iron accumulation in mitochondria. Similar effects have been demonstrated in human disease with increased markers of oxidative DNA damage in urine and impaired oxidative phosphorylation in *in vivo* exercise studies using 31 Phosphorus magnetic resonance spectroscopy (^{31}P-MRS). Therapeutical trials mainly focus on antioxidative treatment with coenzyme Q10 or its short-chain variant idebenone. Promising effects on cardiac hypertrophy in uncontrolled preliminary studies contrast with minor effects in controlled trials and no effect of antioxidants on neurological deficits has been established. Preliminary encouraging ^{31}P-MRS data exist for the treatment with L-carnitine but not with creatine. However, all these interventions may take effect too late in the pathogenic process. Alternative strategies aiming at an enhancement of frataxin by stem cell transplantation, gene transfer or frataxin supplementation are desirable. Additionally, more efficient biomarkers are needed to monitor treatment effects.

Introduction

In 1863 Nikolaus Friedreich published his first of five essays on the disease that later bore his name. He described a familial disorder starting about puberty and causing progressive afferent ataxia due to degeneration of dorsal columns of the spinal cord with loss of muscle stretch reflexes. Additional degeneration was found in corticospinal tracts causing weakness and extensor plantar responses. Dysarthria developed in later stages of the disease. Friedreich also recognized skeletal deformities like pes cavus and thoracic scoliosis as well as cardiomyopathy as part of the disease. In his remarks on therapy Friedreich states that cod-liver oil, iodine iron, nux vomica and argentum nitricum had absolutely no effect. Similarly, application of electric current, courses of therapeutic baths and hot irons at both sides of the spinal column were discouraging because of slow but clear progression despite all these therapies whereas apparent stabilisation occurred especially in times when the patients were without any treatment apart from diet (Friedreich, 1863).

From clinical description to definite genetic diagnosis

Imprecise use of the term Friedreich's ataxia (FA) led to confusion and hampered progress for many decades. In 1981 Harding set up thorough diagnostic criteria which she regarded as obligate for the diagnosis of Friedreich's ataxia: 1) Progressive ataxia, 2) Early onset of symptoms before 25 years of age, 3) Family history compatible with autosomal recessive disease or sporadic occurrence, 4) Absence of knee and ankle jerks due to 5) Sensory axonal neuronopathy, 6) Extensor plantar responses and 7) Dysarthria sometimes developing late in the course of the disease (Harding, 1981). This strict description was essential for the collection of a genetically homogeneous sample of patients and families that in 1996 helped to detect the responsible gene on chromosome 9. FA was the first autosomal recessive disease found to be caused by trinucleotide repeat expansions. 98% of FA chromosomes carry an expanded GAA repeat in the first intron of the *FRDA gene*. Whereas normal alleles carry 7–29 GAA motifs, the repeat is expanded to 66–1300 GAA motifs in FA patients (Campuzano et al., 1996; Epplen et al., 1997). The remaining 2% of FA chromosomes show diverse truncating and mis-sense mutations (Campuzano et al., 1996). Since FA patients with two point mutations have not been identified yet, determination of the GAA repeat length is sufficient for a definite diagnosis. Genetic analyses proved that almost all patients fulfilling the diagnostic criteria established by Harding carried the repeat expansion. However, about 20% of genetically confirmed FA patients show atypical phenotypes. Onset may be as late as the seventh decade, reflexes can be hyperactive, plantar responses may remain normal and in single patients even sensory nerve action potentials are not decreased. Genetically, patients with atypical phenotypes carry relatively mild expansions or mis-sense mutations (D112Y, G130V) in the amino terminal half of the mature *FRDA gene* product frataxin (Durr et al., 1996; Schöls et al., 1997; Cossee et al., 1999).

Pathogenesis in FA models

Repeat expansions as well as point mutations in the *FRDA gene* reduce expression of frataxin (Campuzano et al., 1997; Bidichandani et al., 1998). Frataxin is a nuclear encoded protein localised at mitochondrial membranes and crests. Expression studies demonstrated high levels of frataxin in major sites of FA pathology like dorsal root ganglia where loss of sensory neurons causes sensory axonal neuropathy and degeneration of the posterior columns as well as secondary affection of spinocerebellar tracts and Clark's columns (Campuzano et al., 1996). High frataxin expression in non-neuronal tissue such as heart and pancreas may well account for cardiomyopathy and an increased incidence of diabetes observed in FA patients. Despite the physiological function of frataxin

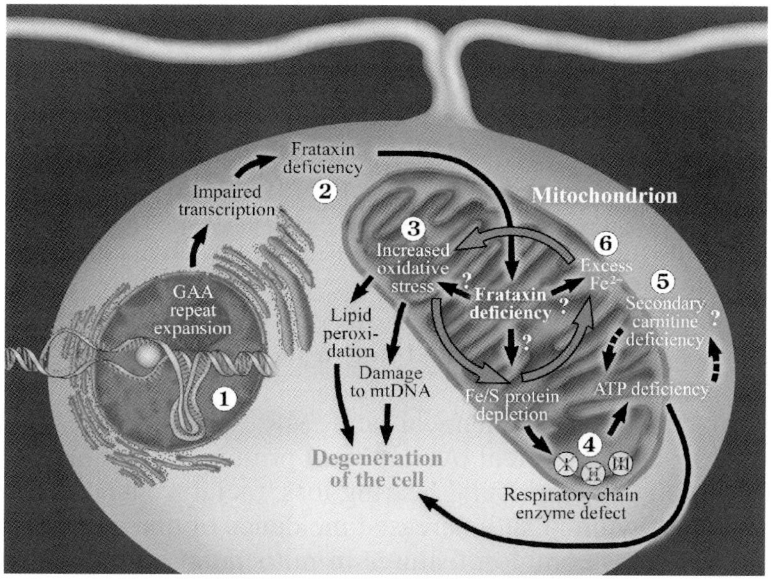

Fig. 1. Pathogenic concepts in Friedreich's ataxia. Mutations cause reduced levels of frataxin resulting in increased oxidative stress, impaired iron-sulphur-(Fe/S)-cluster assembly and consequent depletion of respiratory chain complexes I–III as well as aconitase. Numbers indicate potential therapeutic interventions: 1) Gene transfer, 2) frataxin supplementation, 3) antioxidants, 4) enhancers of mitochondrial energy metabolism, 5) compensation of secondary carnitine deficiency and 6) mitochondrial iron chelators

is not yet known some indications derive from yeast and animal models. Frataxin is a highly preserved protein throughout evolution. Homologue genes occur in yeast as well as in rodents. Yeast strains in which the frataxin homologue had been destroyed show increased mitochondrial iron, increased sensitivity to oxidative stress, and impaired respiratory chain function (Babcock et al., 1997; Wilson et al., 1997; Foury et al., 1997). Mice lacking the frataxin homologue show early embryonic death without accumulation of iron (Cossee et al., 2000). Conditional knock-out mice lacking the frataxin homologue in neurons and cardiac muscle or in skeletal muscle develop severe disease with impaired oxidative phosphorylation due to lack of iron-sulphur-cluster dependent proteins like respiratory chain complexes I–III and aconitases. Iron accumulation occurs after onset of pathology and secondary to impaired respiratory chain function (Puccio et al., 2001). These observations suggest a role for frataxin in iron-sulphur-cluster assembly, as a mitochondrial antioxidant or as a regulating factor in iron homeostasis (Fig. 1).

Friedreich's ataxia a mitochondrial disease?

Several lines of evidence suggest similar mechanisms to be involved in animal models and human disease. In cardiac biopsies from FA patients activities of aconitase and respiratory chain complexes I–III are substantially decreased (Rötig et al., 1997). A DNA marker of oxidative stress (8OH2′dG) was found to be increased in urine of FA patients (Schulz et al., 2000). Oxidative phosphorylation assessed by phosphorus magnetic resonance spectroscopy (^{31}P-MRS) in skeletal muscle demonstrated prolonged phosphocreatine recovery as a direct measure of impaired ATP formation in FA patients *in vivo*. In correspondence, cardiac ^{31}P-MRS found a reduced phosphocreatine to ATP ratio as a measure of impaired cardiac energetic state (Lodi et al., 1999; Vorgerd et al., 2000; Lodi et al., 2001).

Iron accumulation has been described in cardiac muscle (Lamarche et al., 1980) and fibroblasts (Delatycki et al., 1999). In the dentate nucleus of FA patients signal reduction is seen on MRI but this may reflect the physical state of iron shifted to glial cells of the atrophic nucleus (Waldvogel et al., 1999; Koeppen, personal communication). However, iron deposits have not been observed in other autopsy studies and not in most affected tissues like dorsal root ganglia.

In concordance with biochemical and MR spectroscopic findings, FA patients show substantial clinical overlap with primary mitochondrial diseases. Ataxia, optic atrophy, sensorineural hearing loss, facial weakness, hypertrophic cardiomyopathy, arrhythmia and increased incidence of diabetes are frequent in FA and are regarded as common features in mitochondrial cytopathies.

From pathogenesis to therapy

On the background of the pathogenetic concepts derived from animal models as well as pathoanatomical, biochemical and MR spectroscopic studies in human disease new concepts have been developed for the treatment of FA (Table 1, Fig. 1).

Defective respiratory chain complexes and their sensitivity to oxidative stress suggest antioxidants and agents enhancing mitochondrial energy metabolism as promising compounds for the treatment of FA. A natural antioxidant and essential part of the respiratory chain is the quinone coenzyme Q10. Its analogue, idebenone, is more ready to cross membranes including the blood-

Table 1. Therapeutic options in Friedreich's ataxia

Intervention	Potential effect
Stem cell transplantation	Restitution of e.g. pseudounipolar sensory ganglia cells
Gene transfer	Transfer of a normal copy of the *FRDA gene* e.g. to residual pseudounipolar sensory ganglia cells
Recombinant frataxin	Supplementation e.g. of residual pseudounipolar sensory ganglia cells with frataxin may restore normal function
Antioxidants	E.g. coenzyme Q10, idebenone or vitamin E may substitute the supposed antioxidative potential of frataxin and protect respiratory chain complexes
Enhancers of energy metabolism	L-carnitine may compensate potential secondary carnitine deficiency. Creatine may function as intrinsic energy reservoir to regenerate ATP
Iron chelators	Mitochondrial Fe^{2+} chelators e.g. PCIH may decrease mitochondrial iron and oxidative stress

PCIH 2-Pyridylcarboxaldehyde isonicotinoyl hydrazone

brain-barrier because of a shortened side chain. In an *in vitro* assay idebenone (in the presence of succinate) prohibited the toxic effects of ferrous iron to respiratory chain activity in cardiac muscle homogenate (Rustin et al., 1999). Indeed, the protective effect of idebenone on cardiomyopathy was supported by preliminary trials with idebenone (5 mg/kg/d) in FA. In an open trial echocardiography suggested a decrease in left ventricular mass (>20% in 50% of patients) and improved cardiac function but no improvement of ataxia after 6 months of treatment with idebenone (Hausse et al., 2002; Rustin et al., 2002). Other groups could not confirm cardiac effects but found improved performance in the ataxia rating scale after 3–12 months on idebenone (Artuch et al., 2002). Placebo-controlled trials could not prove an effect on ataxia or oxidative phosphorylation assessed by ^{31}P-MRS but found a small but significant effect (4.6% left ventricular mass reduction) on cardiac hypertrophy (Schöls et al., 2001; Mariotti et al., 2003). However, in young FA patients (mean age 15 years), idebenone could not halt the progression of ataxia over 1 year of treatment (Buyse et al., 2003). Coenzyme Q10 (400 mg/d) in combination with vitamin E (2.100 IE/d) is reported to improve cardiac energy metabolism in ^{31}P-MRS in an open trial. No effect on left ventricular mass or ataxia had been found (Lodi et al., 2001). All idebenone and coenzyme Q10 studies in FA reported so far are summarized in Table 2.

Contradictory results may result from different study designs (uncontrolled versus placebo-control) or depend on age of patients (adult versus juvenile), duration of study phases (6 weeks to 1 year) and different assessment of cardiac and neurological function. In summary, the question whether Idebenone, coenzyme Q10 or vitamin E is beneficial in FA remains open and requires larger placebo-controlled trials with longer study periods and sensitive progression markers. Furthermore, no dosage ascertainment has been performed hitherto.

Transgenic mice do not respond to idebenone up to 30 mg/kg/d. Only with 90 mg/kg/d a mild but significant improvement of survival (10%) could be reached (M. Koenig, personal communication). The present impression is that antioxidants are of limited effects in FA treatment and search for cure or halting disease progression requires additional strategies. However, only very few studies faced alternative substances.

L-carnitine may improve mitochondrial function in FA by compensating carnitine deficiency that potentially derives secondary to impaired OXPHOS. Defective OXPHOS with increased NADH/NAD(+) ratio inhibits β-oxidation and produces secondary carnitine deficiency while increasing reactive oxygen species and depleting alpha-tocopherol (Infante and Huszagh, 2000). In a placebo-controlled triple-phase crossover trial the effects of L-carnitine (3 g/d) and creatine (6.75 g/d), two compounds that potentially enhance mitochondrial energy metabolism, were studied over a period of 4 months each. L-carnitine but not creatine improved oxidative phosphorylation in ^{31}P-MRS compared to baseline. However, the effect lost significance in comparison to placebo ($p = 0.06$). No effects were observed concerning ataxia or cardiac hypertrophy (Schöls et al., 2002). Another trial with L-acetylcarnitine in FA patients demonstrated mild improvement in coordination assessed by clinical scores after 3 and

Table 2. Idebenone and coenzyme Q10 trials in Friedreich's ataxia

Substance (Dosage)	No. of patients (Age)	Study period [months]	Control group	Effect on LV mass	Effect on ataxia	Other effects	Reference
Idebenone (5 mg/kg/d)	3 (11–21)	4–9	No	Reduction by 21–36%	No		Rustin (1999)
Idebenone (360 mg/d)	9 (19–54)	1.5	Placebo	No	No	No effect on energy metabolism in 31P-MRS	Schöls (2001)
Idebenone (5 mg/kg/d)	38 (4–22)	6	No	Reduction of >20% in 50% of patients	No	Shortening fraction improved in 5 of 6 pats	Rustin (2002) Hausse (2002)
Idebenone (5 mg/kg/d)	9 (11–19)	12	No	No	Improved after 3–12 months		Artuch (2002)
Idebenone (5 mg/kg/d)	29 (20–31)	12	Placebo	4.6% reduction	No	Only patients with IVS hypertrophy enclosed	Mariotte (2003)
Idebenone (5 mg/kg/d)	8 (8–27)	12	No	16% reduction	Progressive deterioration		Buyse (2003)
Coenzyme Q10 (400 mg/d) + Vitamin E (2100 IE/d)	10 (16–40)	6	No	No	No	Cardiac energy metabolism improved in 31P-MRS	Lodi (2001)

IVS Interventricular septum; *31P-MRS* Phosphorus magnetic resonance spectroscopy

6 months with L-acetylcarnitine but the effects were not compared to placebo (Sorbi et al., 2000).

Iron chelators have been proposed to reduce iron transmitted oxidative stress. However, iron chelators like desferroxamine deplete cytoplasmic but not mitochondrial iron and may worsen the disease. New iron chelators designed to reduce mitochondrial iron like 2-pyridylcarboxaldehyde isonicotinoyl hydrazone (PCIH) and derivates may overcome this problem but their effectiveness has to be questioned given the few observations of increased iron in FA tissues and the late accumulation of iron clearly secondary to impaired respiratory chain function in transgenic mice (Puccio et al., 2001; Richardson, 2003).

Treatment within reach?

The previous experiences of therapeutical trials in FA require new strategies to develop an effective treatment in FA:

- Alternative pathogenic concepts for FA should be taken into account considering frataxin function in iron-sulphur-cluster assembly (Tan et al., 2003), preservation of aconitase activity and activation of mitochondrial energy conversion (Ristow et al., 2000).
- New compounds should be tested either deriving from new pathogenic concepts or with more specific antioxidative function in the mitochondrion.
- Before entering clinical trials, substances should show promising results in animal models.
- Knowledge about the natural course of FA has to be improved. Additionally, it remains a major challenge to establish sensitive and highly reproducible progression markers for the assessment of cardiac and neurological function to facilitate clinical trials with realistic numbers of patients and realistic study periods in the near future.
- On this background the concentration of recent studies on left ventricular (LV) mass assessed by echocardiography as the primary outcome measure has to be questioned. Echocardiography calculates LV mass using formula with cubes of two focal measurements of wall thickness (Interventricular septum, posterior wall) and the enddiastolic diameter (Devereux et al., 1984). LV mass assessed by echocardiography shows a poor interrater variability of 25% substantially decreasing power. Cardiac MRI (Fig. 2) is a much more reliable measure of left ventricular hypertrophy with an interrater variability of less than 5%. Additionally, only a limited number of adult FA patients present with increased left ventricular mass. No reliable long-term data exist about the natural course of cardiac hypertrophy. Limited retrospective echocardiographic data from our series showed spontaneous decrease of cardiac hypertrophy especially in young patients and highly variable cardiac function even without medical intervention. This underscores that placebo controls are essential even in apparently chronic progressive diseases like FA.

However, all these interventions may take effect too late in the pathogenic process. Gene therapy transferring a normal copy of the *FRDA gene* e.g. to dorsal

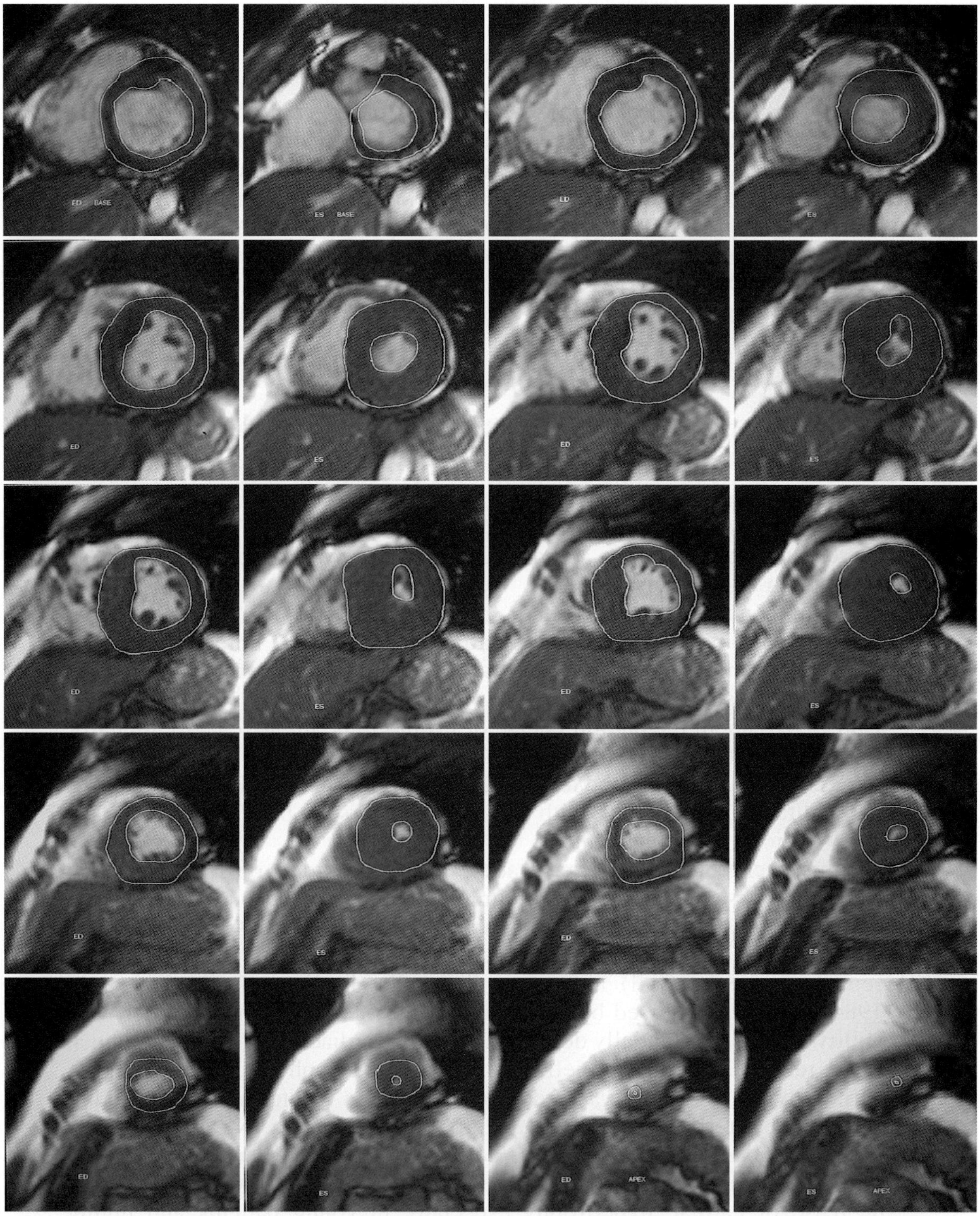

Fig. 2. Assessment of left ventricular mass by cardiac MRI using ECG-triggered cine true-fisp sequences in the short axis plane with 8 mm slices from the mitral valve to the apex. Endocardial and epicardial boundaries of every slice are traced manually. Specific software calculates the enddiastolic volume, endsystolic volume, ejection fraction and left ventricular mass with an interrater variability of less than 5%. Enddiastolic images are presented in the first and third column with corresponding endsystolic images in the second and fourth column, respectively

root ganglia, cardiac myocytes and β-cells of the pancreas may help to express a sufficient amount of functional frataxin in critical tissues. This appears to be not totally out of reach since *FRDA* is a relatively small gene spanning 80 kb of genomic DNA and a major transcript of 1.3 kb encoding a protein of 210 amino acids (Campuzano et al., 1996). Similarly, supplementation with recombinant frataxin could be effective if transfer to neuronal and cardiac tissues can be provided. Alternatively, stem cells may replace degenerating neurons by cells expressing normal levels of frataxin. Stem cell transplantation may be the strategy of choice since severe loss of neurons e.g. in dorsal root ganglia can be demonstrated electrophysiologically by massively decreased sensory nerve action potentials already at the onset of clinical symptoms.

Intensive research work on animal models, natural history and progression markers as well as gene transfer and stem cell techniques should help that the title of a recent review becomes true: "Friedreich's ataxia: treatment within reach" (Dürr et al., 2002).

Acknowledgements

We thank H. Schulze for her kind and professional assistence in the scientific illustration of this article. This study has been supported by a grant of the Deutsche Forschungsgemeinschaft (Scho 754/1-1) to LS.

References

Artuch R, Aracil A, Mas A, Colome C, Rissech M, Monros E, Pineda M (2002) Friedreich's ataxia: idebenone treatment in early stage patients. Neuropediatrics 33: 190–193

Babcock M, de Silva D, Oaks R, Davis-Kaplan S, Jiralerspong S, Montermini L, Pandolfo M, Kaplan J (1997) Regulation of mitochondrial iron accumulation by Yfh1p, a putative homolog of frataxin. Science 276: 1709–1712

Bidichandani SI, Ashizawa T, Patel PI (1998) The GAA triplet-repeat expansion in Friedreich ataxia interferes with transcription and may be associated with an unusual DNA structure. Am J Hum Genet 62: 111–121

Buyse G, Mertens L, Di Salvo G, Matthijs I, Weidemann F, Eyskens B, Goossens W, Goemans N, Sutherland GR, Van Hove JL (2003) Idebenone treatment in Friedreich's ataxia: neurological, cardiac, and biochemical monitoring. Neurology 60: 1679–1681

Campuzano V, Montermini L, Moltò MD, et al. (1996) Friedreich's ataxia: autosomal recessive disease caused by an intronic GAA triplet repeat expansion. Science 271: 1423–1427

Campuzano V, Montermini L, Lutz Y, Cova L, Hindelang C, Jiralerspong S, Trottier Y, Kish SJ, Faucheux B, Trouillas P, Authier FJ, Durr A, Mandel JL, Vescovi A, Pandolfo M, Koenig M (1997) Frataxin is reduced in Friedreich ataxia patients and is associated with mitochondrial membranes. Hum Mol Genet 6: 1771–1780

Cossee M, Durr A, Schmitt M, Dahl N, Trouillas P, Allinson P, Kostrzewa M, Nivelon-Chevallier A, Gustavson KH, Kohlschutter A, Muller U, Mandel JL, Brice A, Koenig M, Cavalcanti F, Tammaro A, De Michele G, Filla A, Cocozza S, Labuda M, Montermini L, Poirier J, Pandolfo M (1999) Friedreich's ataxia: point mutations and clinical presentation of compound heterozygotes. Ann Neurol 45: 200–206

Cossee M, Puccio H, Gansmuller A, Koutnikova H, Dierich A, LeMeur M, Fischbeck K, Dolle P, Koenig M (2000) Inactivation of the Friedreich ataxia mouse gene leads to early embryonic lethality without iron accumulation. Hum Mol Genet 9: 1219–1226

Delatycki MB, Camakaris J, Brooks H, Evans-Whipp T, Thorburn DR, Williamson R, Forrest SM (1999) Direct evidence that mitochondrial iron accumulation occurs in Friedreich ataxia. Ann Neurol 45: 673–675

Devereux RB, Reichek N (1977) Echocardiographic determination of left ventricular mass in man. Anatomic validation of the method. Circulation 55: 613–618

Dürr A (2002) Friedreich's ataxia: treatment within reach. Lancet Neurol 1: 370–374

Dürr A, Cossée M, Agid Y, et al. (1996) Clinical and genetic abnormalities in patients with Friedreich's ataxia. N Engl J Med 335: 1169–1175

Epplen C, Epplen JT, Frank G, et al. (1997) Differential stability of the $(GAA)_n$ tract in the Friedreich Ataxia (STM7) gene. Hum Genet 99: 834–836

Foury F, Cazzalini O (1997) Deletion of the yeast homologue of the human gene associated with Friedreich's ataxia elicits iron accumulation in mitochondria. FEBS Lett 411: 373–377

Friedreich N (1863) Über degenerative Atrophie der spinalen Hinterstränge. Virchows Arch Pathol Anat Physiol 26: 391–419

Harding AE (1981) Friedreich's ataxia: a clinical and genetic study of 90 families with an analysis of early diagnostic criteria and intrafamilial clustering of clinical features. Brain 104: 589–620

Hausse AO, Aggoun Y, Bonnet D, et al. (2002) Idebenone and reduced cardiac hypertrophy in Friedreich's ataxia. Heart 87: 346–349

Infanate JP, Huszagh VA (2000) Secondary carnitine deficiency and impaired docosahexaenoic (22:6n-3) acid synthesis: a common denominator in the pathophysiology of diseases of oxidative phosphorylation and beta-oxidation. FEBS Lett 468: 1–5

Lamarche JB, Cote M, Lemieux B (1980) The cardiomyopathy of Friedreich's ataxia – morphological observations in 3 cases. Can J Neurol Sci 7: 389–396

Lodi R, Cooper JM, Bradley JL, et al. (1999) Deficit of in vivo mitochondrial ATP production in Patients with Friedreich's ataxia. Proc Natl Acad Sci USA 96: 11492–11495

Lodi R, Hart PE, Rajagopalan B, et al. (2001) Antioxidant treatment improves in vivo cardiac and skeletal muscle bioenergetics in patients with Friedreich's ataxia. Ann Neurol 49: 590–596

Mariotti C, Solari A, Torta D, Marano L, Fiorentini C, Di Donato S (2003) Idebenone treatment in Friedreich patients: one-year-long randomized placebo-controlled trial. Neurology 60: 1676–1679

Puccio H, Simon D, Cossee M, Criqui-Filipe P, Tiziano F, Melki J, Hindelang C, Matyas R, Rustin P, Koenig M (2001) Mouse models for Friedreich ataxia exhibit cardiomyopathy, sensory nerve defect and Fe-S enzyme deficiency followed by intramitochondrial iron deposits. Nat Genet 27: 181–186

Richardson DR (2003) Friedreich's ataxia: iron chelators that target the mitochondrion as a therapeutic strategy? Expert Opin Invest Drugs 12: 235–245

Ristow M, Pfister MF, Yee AJ, Schubert M, Michael L, Zhang CY, Ueki K, Michael MD 2nd, Lowell BB, Kahn CR (2000) Frataxin activates mitochondrial energy conversion and oxidative phosphorylation. Proc Natl Acad Sci USA 97: 12239–12243

Rötig A, de Lonlay P, Chretien D, Foury F, Koenig M, Sidi D, Munnich A, Rustin P (1997) Aconitase and mitochondrial iron-sulphur protein deficiency in Friedreich ataxia. Nat Genet 17: 215–217

Rustin P, von Kleist-Retzow JC, Chantrel-Groussard K, Sidi D, Munnich A, Rotig A (1999) Effect of idebenone on cardiomyopathy in Friedreich's ataxia: a preliminary study. Lancet 354: 477–479

Rustin P, Rotig A, Munnich A, Sidi D (2002) Heart hypertrophy and function are improved by idebenone in Friedreich's ataxia. Free Radic Res 36: 467–469

Schöls L, Amoiridis G, Przuntek H, et al. (1997) Friedreich's ataxia: revision of the phenotype according to molecular genetics. Brain 120: 2131–2140

Schöls L, Vorgerd M, Schillings M, Skipka G, Zange J (2001) Idebenone in patients with Friedreich ataxia. Neurosci Lett 306: 169–172

Schöls L, Vorgerd M, Abele M, Schillings M, Skipka G, Müller K, Klockgether T, Przuntek H, Zange J (2002) A placebo-controlled crossover trial of carnitine and creatine in Friedreich's ataxia. J Neurol 249 [Suppl I]: 56

Schulz J, Dehmer T, Schöls L, Mende H, Hardt C, Vorgerd M, Bürk K, Dichgans J, Beal MF, Bogdanov MB (2000) Oxidative stress in Friedreich ataxia. Neurology 55: 1719–1721

Sorbi S, Forleo P, Fani C, Piacentini S (2000) Double-blind, crossover, placebo-controlled clinical trial with L-acetylcarnitine in patients with degenerative cerebellar ataxias. Clin Neuropharmacol 23: 114–118
Tan G, Napoli E, Taroni F, Cortopassi G (2003) Decreased expression of genes involved in sulfur amino acid metabolism in frataxin-deficient cells. Hum Mol Genet 12: 1699–1711
Vorgerd M, Schöls L, Hardt C, Ristow M, Epplen JT, Zange J (2000) Mitochondrial impairment of human muscle in Friedreich ataxia *in vivo*. Neuromuscular Disord 10: 430–435
Waldvogel D, van Gelderen P, Hallett M (1999) Increased iron in the dentate nucleus of patients with Friedrich's ataxia. Ann Neurol 46: 123–125
Wilson RB, Roof DM (1997) Respiratory deficiency due to loss of mitochondrial DNA in yeast lacking the frataxin homologue. Nat Genet 16: 352–357

Authors' address: Prof. L. Schöls, Center of Neurology and Hertie-Institute for Clinical Brain Research, University of Tübingen, Hoppe-Seyler-Strasse 3, D-72076 Tübingen, Germany, e-mail: Ludger.Schoels@uni-tuebingen.de

J Neural Transm (2004) [Suppl] 68: 147–156

Cholecystokinin- and cholecystokinin-B-receptor gene polymorphisms in panic disorder

V. G. Hösing[1], **A. Schirmacher**[6], **G. Kuhlenbäumer**[6], **C. Freitag**[8], **P. Sand**[2], **C. Schlesiger**[1], **C. Jacob**[2], **J. Fritze**[4], **P. Franke**[3], **M. Rietschel**[3,5], **H. Garritsen**[7], **M. M. Nöthen**[9], **R. Fimmers**[8], **F. Stögbauer**[6], and **J. Deckert**[1]

[1] Department of Psychiatry, University of Münster,
[2] Department of Psychiatry, University of Würzburg,
[3] Department of Psychiatry, University of Bonn,
[4] Department of Psychiatry, University of Frankfurt,
[5] Central Institute of Mental Health, Mannheim,
[6] Department of Neurology, and
[7] Institute of Transfusion Medicine and Transplantation Immunology, University of Münster, and
[8] Institute of Medical Biometry, Informatics, and Epidemiology, University of Bonn, Germany
[9] Department of Medical Genetics, University of Antwerp, Belgium

Summary. Panic disorder like other neuropsychiatric disorders is believed to be caused by multiple psychosocial and biological factors. Several lines of evidence point to a role for the peptide neurotransmitter cholecystokinin in the pathogenesis of panic disorder. We therefore determined the allele and genotype frequencies of a single nucleotide polymorphism in the CCK gene (−36C > T) and one CT repeat polymorphism in the CCK-B-receptor gene in a German panic disorder sample ($n = 115$ for CCK gene polymorphism, $n = 111$ for CCK-B-receptor polymorphism) and compared them with gender and age matched controls. The length of the polymorphic CT repeat alleles varies between 146 bp and 180 bp. We first analysed the results by a permutation test which provided evidence for heterogeneity between patients and controls ($p = 0.002$). We then analysed the data as a di-allelic polymorphism with a short (146–162 bp) and a long (164–180 bp) allele and as a tetra-allelic polymorphism with 4 alleles (146–154 bp, 156–162 bp, 164–170 bp, 172–180 bp). In the di-allelic analysis as well as in the tetra-allelic analysis there was an excess of the longer allele ($p = 0.001$) or the two longer alleles ($p = 0.041$) respectively in patients with panic disorder. No difference between groups was observed for the −36C > T polymorphism. Our findings are consistent with the notion that genetic variation in the CCK neurotransmitter system contributes to the pathogenesis of panic disorder.

Introduction

The anxiety disorder panic disorder is one of the most common mental disorders with a lifetime prevalence of 1–3% (Weissman et al., 1997). It is characterized by unexpected, recurrent panic attacks and often accompanied by agoraphobia. Twin and family studies indicate that genetic factors are an important factor in the pathogenesis of panic disorder with an estimated heritability of 48% (Hettema et al., 2001; Maier et al., 1993). Segregation analyses did not provide evidence for Mendelian inheritance and were interpreted as consistent with the assumption that the effects of individual genes are minor (Vieland et al., 1996). Accordingly, linkage analyses showed significant results only in genetic isolates such as Iceland (Thorgeirsson et al., 2003) or under the assumption of syndromes combining panic disorder and somatic disorders (Gratacos et al., 2001; Hamilton et al., 2003). Association analyses with candidate genes, which are able to detect genes with minor effects, provided evidence in independent samples for contributions of polymorphisms in the monoamine oxidase A gene, the catechol-O-methyltransferase gene and the gene of the adenosine A_{2A} receptor (Deckert et al., 1998, 1999; Hamilton et al., 2002).

Genes for association studies have been identified on the basis of our knowledge of panic disorder pathophysiology. The gastrin-like neuropeptide cholecystokinin (CCK) (Vancerhaeghen et al., 1975) is thought to play a role in the pathogenesis of panic disorder, because it provokes panic attacks in patients with significantly greater efficacy than in controls (Bradweijn et al., 1991). CCK receptors are differentiated into CCK-A and CCK-B receptors. While the CCK-A receptor is located in the gastrointestinal tract, the CCK-B receptor is essentially restricted to the central nervous system and may therefore be more relevant for mental disorders (Moran et al., 1986). The genes for CCK and the CCK-B-receptor have been characterized and have been located on chromosomal regions 3pter-p21 and 11p15.4, respectively (Takahashi et al., 1986; Song et al., 1993). Contradictory association findings have been reported both for a polymorphism in the CCK gene (−36C>T: positive Wang et al., 1998 versus negative Kennedy et al., 1998; Hattori et al., 2001; Hamilton et al., 2001) as well as in the CCK-B-receptor gene (CT repeat polymorphism: positive Kennedy et al., 1999 versus negative Hamilton et al., 2001; Hattori et al., 2001; Yamada et al., 2001).

Stimulated by these previous studies, we investigated the single nucleotide polymorphism (SNP) in the promotor of the CCK gene (−36C>T) and the CT repeat polymorphism of the CCK-B-receptor gene in a German sample of panic disorder patients in comparison to gender-and age-matched controls.

Materials and methods

A sample of 115 unrelated German patients with panic disorder was investigated in this study (female n = 74, male n = 41; mean age: 37.8 ± 11.0 years). In these patients, panic disorder was diagnosed by experienced psychiatrists on the basis of structured clinical interviews according to the criteria of DSM-III-R or DSM-IV (SADS-LA (Mannuzza et al., 1986) and CIDI (Robins et al., 1988; Wittchen et al., 1997)) and a review of medical records. The control group consisted of 115 unrelated anonymous blood donors of German descent, who were matched according to gender and age (female n = 74, male n = 41; mean age: 41.5 ± 10.6 years).

The study was approved by the respective local ethical committees and informed consent was obtained from all participating subjects.

DNA was extracted from EDTA blood samples by using the QIAamp-DNA MiniKit® (Qiagen GmbH, Hilden, Germany).

PCR reactions for the CCK gene polymorphism were performed in a total volume of 25 µl containing 10 pmol forward primer 5′-GGA TTA ACT CCA CCC CAC TAG AC-3′, 10 pmol reverse primer 5′-ATT AGG ACG CAG CTG GCT TG-3′, 120 ng template DNA, 200 µM dNTPs, 2.5 µl PCR buffer (15 mM $MgCl_2$, Tris-Cl, KCL, $(NH_4)_2SO_4$, Qiagen GmbH, Hilden, Germany), 13.8 µl H_2O and 1 Unit HotstarTaq™ DNA polymerase at 94°C for 10 min, followed by 35 cycles of 94°C for 30 s, 57°C for 30 s, 72°C for 30 s with a final 10 min extension at 72°C in a Gene Amp PCR system 9600 (Applied Biosystems, Weiterstadt, Germany). Genotyping was done by means of restriction fragment length polymorphism (RFLP). PCR products were digested by the restriction enzyme Hin6I at 37°C for 24 hours. 8 µl of the digested product were mixed with 12 µl denaturing solution and separated for 3 h on a 15% polyacrylamide gel (acrylamide:bisacrylamide = 49:1; Multigel-Long/Biometra, Göttingen, Germany) containing 1 × TBE at 20 V/cm. Bands were visualized by silver-staining. PCR reactions for the CCK-B receptor CT VNTR polymorphism were performed in a total volume of 25 µl containing 10 pmol forward primer (5′-TCT ACC CAA CCC CTC CTA CC-3′, 5′ FAM fluorescently labeled), 10 pmol reverse primer (5′-CAC CTC CTC TTT ATC ACC CGC-3′), 120 ng template DNA, 200 µM dNTPs, 2.5 µl PCR buffer (15 mM $MgCl_2$, Tris-Cl, KCL, $(NH_4)_2SO_4$, Qiagen, GmbH, Hilden, Germany), 13.8 µl H_2O and 1 Unit HotstarTaq™ DNA polymerase and were cycled at 94°C for 10 min followed by 35 cycles of 94°C for 30 s, 60°C for 30 s, 72°C for 30 s with a final 10 min at 72°C. Genotyping was done by means of fragment analysis using the 377 Applied Biosystems automatic sequencer (Applied Biosystems, Weiterstadt, Germany). Fragment sizes varied between 146 bp and 180 bp.

The case-control setting of the program FASTEHPLUS (Zhao et al., 2000) was used to asses differences in the genotype distribution between cases and controls. FASTEHPLUS performs model-free analysis and permutation tests of allelic association via a likelihood ratio test statistic which compares the likelihood of the data in cases and controls versus the likelihood in cases and the likelihood in controls. We performed 10 000 permutations of the distribution of the CCK-B receptor CT VNTR polymorphism genotypes in cases and controls. For the comparison of the di- and tetra-allelic genotypes between cases and controls, χ^2-Tests for alleles and Armitage-Trend- or Fisher's-Exact-Tests for genotypes were performed (Sasieni, 1997). The CCK-B receptor CT VNTR polymorphism alleles were comprised either in 2 categories (short allele 146–162 bp; long allele 164–180 bp) or into 4 categories (146–154 bp; 156–162 bp; 164–170 bp; 172–180 bp). To assess the effect of single alleles we first regressed each allele individually and, second, combined the relevant alleles in a logistic regression. For the logistic regression we combined the homozygous and the heterozygous genotypes of each allele because of the small number of homozygous genotypes. These statistics were calculated with the SAS statistical package (SAS/STAT, version 8.1, Cary, NC: SAS Institute Inc., 1999). Hardy-Weinberg equilibrium was examined by the program Finetti (Wienker TF, personal communication).

Results

Hardy-Weinberg equilibrium was observed both for the −36C > T CCK gene polymorphism and the CCK-B-receptor gene polymorphism under di-allelic as well as tetra-allelic analysis in controls as well as in patients (data not shown).

No significant difference was observed between patients and controls for the −36C > T CCK gene polymorphism in the total as well as in subsamples (Table 1).

The likelihood ratio test statistic of the CCK-B-receptor gene CT polymorphism, however, provided evidence for a different distribution of the genotypes between patients and controls. This could be observed for the total sample ($p = 0.002$) as well as for the panic disorder with agoraphobia subsample

Table 1. Genotype and allele frequencies of the CCK gene −36C>T polymorphism in patients with panic disorder (PD) and controls

	Genotypes				Alleles			
	Cases		Controls		Cases		Controls	
		n (percent)		n (percent)		n (percent)		n (percent)
PD combined	C/C	85 (74%)	C/C	92 (80%)	C	198 (86%)	C	206 (90%)
	C/T	28 (24%)	C/T	22 (19%)	T	32 (14%)	T	24 (10%)
	T/T	2 (2%)	T/T	1 (1%)				
	Fisher's exact test: $p = 0.50$				$\chi^2 = 1.443$; 1 DF; $p = 0.25$			
PD with Agoraphobia	C/C	58 (72%)	C/C	92 (80%)	C	136 (85%)	C	206 (90%)
	C/T	20 (25%)	C/T	22 (19%)	T	24 (15%)	T	24 (10%)
	T/T	2 (3%)	T/T	1 (1%)				
	Fisher's exact test: $p = 0.34$				$\chi^2 = 1.82$; 1 DF; $p = 0.18$			

Table 2. Genotype frequencies of the CCK-B-receptor gene CT-repeat polymorphism in patients with panic disorder (PD) and controls

	Tetra-allelic distribution										Di-allelic distribution		
	s/s	s/m	s/l	s/xl	m/m	m/l	m/xl	l/l	l/xl	xl/xl	s/s	s/l	l/l
PD combined n = 111 (percent)	1 (1%)	8 (7%)	22 (20%)	2 (2%)	6 (5%)	20 (18%)	2 (2%)	35 (31%)	15 (14%)	0 (0%)	15 (14%)	46 (41%)	50 (45%)
	Statistic compared to controls: Armitage-Trend-Test[a], Z = −2.04, p = 0.04										Z = −3.30, p = 0.001		
Females n = 71 (percent)	0 (0%)	4 (5%)	15 (21%)	2 (3%)	5 (7%)	10 (14%)	2 (3%)	24 (34%)	9 (13%)	0 (0%)	9 (13%)	29 (41%)	33 (46%)
	Statistic compared to controls: Armitage-Trend-Test, Z = −2.20, p = 0.03 Fisher's exact Test p = 0.03										Z = −3.22, p = 0.001		
Males n = 40 (percent)	1 (2%)	4 (10%)	7 (18%)	0 (0%)	1 (2%)	10 (25%)	0 (0%)	11 (28%)	6 (15%)	0 (0%)	6 (15%)	17 (42.5%)	17 (42.5%)
	Statistic compared to controls: Armitage-Trend-Test, Z = −0.48, p = 0.63 Fisher's exact Test p = 0.12										Z = −1.21, p = 0.23		
PD with Agoraphobia n = 81 (percent)	0 (0%)	6 (7%)	15 (19%)	1 (1%)	6 (7%)	14 (18%)	1 (1%)	26 (32%)	12 (15%)	0 (0%)	12 (15%)	31 (38%)	38 (47%)
	Statistic compared to controls: Armitage-Trend-Test, Z = −2.24, p = 0.03 Fisher's exact Test p = 0.04										Z = −3.04, p = 0.002		
Females n = 56 (percent)	0 (0%)	3 (5%)	12 (21%)	1 (2%)	5 (9%)	9 (16%)	1 (2%)	19 (34%)	6 (11%)	0 (0%)	8 (14%)	23 (41%)	25 (45%)
	Statistic compared to controls: Armitage-Trend-Test, Z = −1.94, p = 0.05 Fisher's exact Test p = 0.10										Z = −2.76, p = 0.006		
Males n = 25 (percent)	0 (0%)	3 (12%)	3 (12%)	0 (0%)	1 (4%)	5 (20%)	0 (0%)	7 (28%)	6 (24%)	0 (0%)	4 (16%)	8 (32%)	13 (52%)
	Statistic compared to controls: Armitage-Trend-Test, Z = −1.25, p = 0.21 Fisher's exact Test p = 0.15										Z = −1.47, p = 0.14		

(*continued*)

Table 2 (*continued*)

	Tetra-allelic distribution										Di-allelic distribution		
	s/s	s/m	s/l	s/xl	m/m	m/l	m/xl	l/l	l/xl	xl/xl	s/s	s/l	l/l
Controls n = 111 (percent)	0 (0%)	13 (12%)	19 (17%)	3 (3%)	17 (15%)	26 (23%)	4 (4%)	25 (22%)	4 (4%)	0 (0%)	30 (27%)	52 (47%)	29 (26%)
Females n = 71 (percent)	0 (0%)	11 (15%)	11 (15%)	2 (3%)	9 (13%)	19 (27%)	3 (4%)	14 (20%)	2 (3%)	0 (0%)	20 (28%)	35 (49%)	16 (23%)
Males n = 40 (percent)	0 (0%)	2 (5%)	8 (20%)	1 (2%)	8 (20%)	7 (18%)	1 (2%)	11 (28%)	2 (5%)	0 (0%)	10 (25%)	17 (43%)	13 (32%)

[a] Due to the vast number of permutations, Fisher's exact Test could not be calculated by SAS. Tetra-allelic distribution: s = alleles 146–154, m = alleles 156–162, l = alleles 164–170, xl = alleles 172–180. Di-allelic distribution: s = alleles 146–162, l = alleles 164–180

($p = 0.006$; $n = 81$) and the panic disorder without agoraphobia subsample ($p = 0.025$; $n = 30$). The observed difference was more due to the women ($p = 0.003$; $n = 71$) than to the men ($p = 0.36$; $n = 40$).

In the di-allelic analysis (Table 2) this corresponded to an excess of the long allele (164–180 bp) in panic disorder ($p = 0.001$ for all patients ($n = 111$), $p = 0.002$ for patients with panic disorder with agoraphobia ($n = 81$)). In the tetra-allelic analysis (Table 2) a smaller, but still significant excess of the two longer alleles in panic disorder could be detected ($p = 0.04$ for all patients, $p = 0.03$ for patients with panic disorder with agoraphobia).

When we assessed the effect of single alleles, we found an elevated odds ratio for allele 166 (OR 2.8, 95%-confidence interval 1.6–5.1) and for allele 168 (OR 3.5, 95% confidence interval 1.7–7.4) combined. In the agoraphobia subsample, we observed similar effects (allele 166: OR 2.9, 95% confidence interval 1.5–5.5; allele 168: OR 3.5, 95% confidence interval 1.6–7.8).

Discussion

Our results contribute to the inconsistent observations reported for an association of cholecystokinin neurotransmitter system gene polymorphisms with panic disorder. While we observed a significant association of the long alleles of the CCK-B-receptor CT repeat polymorphism, we could not detect an association of the CCK gene −36C>T polymorphism with panic disorder.

Our positive findings with regard to the CCK-B-receptor gene CT polymorphism replicate the association reported by Kennedy et al. (1999), which Hamilton et al. (2001), Hattori et al. (2001) and Yamada et al. (2001) were not able to do. Similar as in the study of Kennedy and coauthors (1999) in particular two alleles (166 and 168) are responsible for the observed association. Our negative findings with regard to the −36C>T polymorphism in the CCK gene in agreement with the results of Kennedy et al. (1999), Hattori et al. (2001) and Hamilton et al. (2001) fail to confirm the positive findings of Wang et al. (1998).

There may be several reasons for the observed differences between studies. First, the gender ratio of our sample was similar to that of Kennedy et al. (1999), but different from that of Hamilton et al. (2001) and Hattori et al. (2001). The latter ones included fewer women in their samples. This may be relevant because in our sample the observed effect is mainly due to women, and less to men. Differences in diagnostic criteria and differential comorbidity with agoraphobia and other disorders in the investigated samples may also affect results. In our sample, we observed a considerable rate of comorbidity with agoraphobia and affective disorders, comparable to the rate described in epidemiological studies (Sand et al., 2002). A comparison with the other samples is not possible though because they do not give information of the comorbidity rate. Ethnic differences between study samples may also explain differences. However, our observations with regard to the CCK-B-receptor gene CT polymorphism are in line with those of another mainly Caucasian population, while the −36C>T CCK gene polymorphism results are more in line with those of an Asian population among others.

Another possible explanation for conflicting results may be population-specific linkage disequilibrium with other clinically relevant polymorphisms in the investigated genes. Stratification effects also cannot be completely excluded because in our case-control design we used population-based controls and not family-based controls. In addition, our sample size may have been too small to detect a minor contribution of the −36C>T polymorphism. Finally, the statistical analyses employed for analyzing the CT repeat have been differentially conservative. In particular, the power of linkage analyses and family-based association samples (Hamilton et al., 2001) was too weak to detect minor effects in cases of complex inheritance. In addition, the CLUMP analysis (Hamilton et al., 2001) as opposed to the FASTEHPLUS analysis may underestimate the contribution of rare alleles. On the other hand, the tetra-allelic analysis as well as the di-allelic analysis, which increase the power of the sample by lowering the degrees of freedom, are not based on functional data and their clinical relevance therefore needs to be established.

These issues have to be addressed in future studies with larger samples and family-based controls. These studies should also take into account gender and comorbidity and employ analysis of several polymorphisms per gene including a haplotype-analysis. Positive association findings with such further polymorphisms have already been reported (Hattori et al., 2001; Ebihara et al., 2003). Additional studies are necessary to evaluate the functional consequences of the polymorphic CCK-B-receptor gene CT repeat alleles, in particular alleles 166 and 168.

While CCK-B-receptor antagonists have been demonstrated to be effective antipanic agents in animals (Jenck et al., 1996), so far efforts to show efficacy in patients have failed (van Megen et al., 1997). If the observed association between the CCK-B-receptor gene CT repeat polymorphism and panic disorder is confirmed in further independent samples, CCK-B-receptor antagonists should be re-evaluated as a potential novel therapeutic approach in panic disorder.

In summary, our results provide further evidence for a contribution of genetic variations in the CCK neurotransmitter system, in particular the CCK-B-receptor gene on chromosome 11p15.4 to the pathogenesis of panic disorder.

Acknowledgements

We gratefully acknowledge the skillful technical support of K. Weiss. This research was supported by an IMF grant (DE219915), grants from Deutsche Forschungsgemeinschaft (DE357/2-1 and 2-2, KU1194/2-1) and the BMBF (IZKF 01 KS 9603).

References

Bradweijn J, Koszycki D, Shriqui C (1991) Enhanced sensitivity to cholecystokinin- tetrapeptide in panic disorder: clinical and behavioural findings. Arch Gen Psychiatry 48: 603–610

Deckert J, Catalano M, Syagailo YV, Bosi M, Okladnova O, Di Bella D, Nothen MM, Maffei P, Franke P, Fritze J, Maier W, Propping P, Beckmann H, Bellodi L, Lesch KP (1999) Excess of high activity monoamine oxidase A gene promoter alleles in female patients with panic disorder. Hum Mol Genet 8: 621–624

Ebihara M, Ohba H, Hattoro E, Yamada K, Yoshikawa T (2003) Transcriptional activities of cholecystokinin promoter haplotypes and their relevance to panic disorder susceptibility. Am J Med Genet 118B: 32–35

Gratacos M, Nadal M, Martin-Santos R, Pujana MA, Gago J, Peral B, Armengol L, Ponsa I, Miro R, Bulbena A, Estivill X (2001) A polymorphic genomic duplication on human chromosome 15 is a susceptibility factor for panic and phobic disorders. Cell 106: 367–379

Hamilton SP, Slager SL, Helleby L, Heiman GA, Klein DF, Hodge SE, Weissman MM, Fyer AJ, Knowles JA (2001) No association or linkage between polymorphisms in the genes encoding cholezystokinin and the cholezystokinin B receptor and panic disorder. Mol Psychiatry 6: 59–65

Hamilton SP, Slager SL, Heiman GA, Deng Z, Haghighi F, Klein DF, Hodge SE, Weissman MM, Fyer AJ, Knowles JA (2002) Evidence for a susceptibility locus for panic disorder near the catechol-O-methyltransferase gene on chromosome 22. Biol Psychiatry 51: 591–601

Hamilton SP, Fyer AJ, Durner M, Heiman GA, Baisre de Leon A, Hodge SE, Knowles JA, Weissman MM (2003) Further genetic evidence for a panic disorder syndrome mapping to chromosome 13q. PNAS USA 100: 2550–2555

Hattori E, Ebihara M, Yamada K, Ohba H, Shibuya H, Yoshikawa T (2001) Identification of a short tandem repeat stretch in the 5′-upstream region of the cholecystokinin gene, and its association with panic disorder but not with schizophrenia. Mol Psychiatry 6: 465–470

Hettema JM, Neale MC, Kendler KS (2001) A review and meta-analysis of the genetic epidemiology of anxiety disorders. Am J Psychiatry 158: 1568–1578

Jenck F, Martin JR, Moreau J-L (1996) Behavioral effects of CCKB receptor ligands in a validated simulation of panic anxiety in rats. Eur Neuropsychopharmacol 4: 291–298

Kennedy JL, Bradwejn J, Koszycki D, King N, Crowe R, Vincent J, Fourie O (1999) Investigation of cholecystokinin system genes in panic disorder. Mol Psychiatry 4: 284–285

Maier W, Lichtermann D, Minges J, Oehrlein A, Franke P (1993) A controlled family study in panic disorder. J Psychiatr Res 27: 79–87

Mannuzza S, Fyer AJ, Klein DF, Endicott J (1986) Schedule for Affective Disorders and Schizophrenia-Lifetime Version modified for the study of anxiety disorders (SADS-LA): rationale and conceptual development. J Psychiatr Res 20: 317–325

Moran TH, Robinson PH, Goldrich MS, McHugh PR (1986) Two brain cholecystokinin receptors: implications for behavioral actions. Brain Res 362: 175–179

Robins LN, Wing J, Wittchen HU, Helzer JE, Babor TF, Burke J, Farmer A, Jablenski A, Pickens R, Regier DA, Sartorius N, Towle LH (1988) The Composite International Diagnostic Interview. An epidemiologic instrument suitable for use in conjunction with different diagnostic systems and in different cultures. Arch Gen Psychiatry 45: 1069–1077

Sand PG, Schlurmann K, Luckhaus C, Götz M, Stöber G, Lesch KP, Riederer P, Franke P, Maier W, Nöthen M, Propping P, Fritze J, Deckert J (2002) Estrogen receptor 1 gene (ESR 1) variants in panic disorder (Letter). Am J Med Genet 114: 426–428

Sasieni PD (1997) From genotypes to genes: doubling the sample size. Biometrics 53: 1253–1261

Song I, Brown DR, Wiltshire RN, Gantz I, Trent JM, Yamada T (1993) The human gastrin/cholecystokinin type B receptor gene: alternative splice donor site in exon 4 generates two variant mRNAs. Proc Natl Acad Sci USA 90: 9085–9089

Takahashi Y, Fukushige S, Murotsu T, Matsubara K (1986) Structure of the human cholecystokinin gene and its chromosomal location. Gene 50: 353–366

Thorgeirsson TE, Oskarsson H, Desnica N, Kostic JP, Stefansson JG, Kolbeinsson H, Lindal E, Gagunashvili N, Frigge ML, Kong A, Stefansson K, Gulcher JR (2003) Anxiety with panic disorder linked to chromosome 9q in Iceland. Am J Hum Genet 72: 1221–1230

Vancerhaeghen J-J, Signeau JC, Gepts W (1975) New peptide in the vertebrate CNS interacting with antigastrin antibodies. Nature 257: 604–605

van Megen HJGM, Westenberg HGM, den Boer JA, Slaap B, van Es-Radhakishun F, Pande A (1997) The cholecystokinin-B-receptor antagonist CI-988 failed to affect CCK-4 induced symptoms in panic disorder patients. Psychopharmacol 129(3): 243–248

Vieland VJ, Goodman DW, Chapman T, Fyer AJ (1996) New segregation analysis of panic disorder. Am J Med Genet 67: 147–153

Wang Z, Valdes J, Russell N, Zoega T, Crowe R (1998) Possible association of a cholecystokinin promotor polymorphism with panic disorder. Am J Med Genet 81: 228–234

Weissman MM, Bland RC, Canino GJ, Faravelli C, Greenwald S, Hwu HG, Joyce PR, Karam EG, Lee CK, Lellouch J, Lepine JP, Newman SC, Oakley-Browne MA, Rubio-Stipec M, Wells JE, Wickramaratne PJ, Wittchen HU, Yeh EK (1997) The cross-national epidemiology of panic disorder. Arch Gen Psychiatry 54: 305–309

Wittchen HU, Garczynski E, Pfister H (1997) Composite International Diagnostic Interview according to ICD-10 and DSM-IV. Hogrefe, Göttingen

Yamada K, Hattori E, Shimizu M, Sugaya A, Shibuya H, Yoshikawa T (2001) Association studies of the cholezystokinin B receptor and A2a adenosine receptor genes in panic disorder. J Neural Transm 108: 837–848

Zhao JH, Curtis S, Sham PC (2000) Model-free analysis and permutation tests for allelic associations. Hum Hered 50: 133–139

Authors' address: PD Dr. J. Deckert, Department of Psychiatry, University of Münster, Albert-Schweitzer-Strasse 11, D-48149 Münster, Germany, e-mail: deckertj@uni-muenster.de

Springer and the Environment

WE AT SPRINGER FIRMLY BELIEVE THAT AN INTERnational science publisher has a special obligation to the environment, and our corporate policies consistently reflect this conviction.

WE ALSO EXPECT OUR BUSINESS PARTNERS – PRINTERS, paper mills, packaging manufacturers, etc. – to commit themselves to using environmentally friendly materials and production processes.

THE PAPER IN THIS BOOK IS MADE FROM NO-CHLORINE pulp and is acid free, in conformance with international standards for paper permanency.